SpringerBriefs in Applied Sciences and Technology

Thermal Engineering and Applied Science

Series Editor

Francis A. Kulacki, Minnesota, USA

More information about this series at http://www.springer.com/series/8884

Stefan aus der Wiesche • Christian Helcig

Convective Heat Transfer From Rotating Disks Subjected To Streams Of Air

 Springer

Stefan aus der Wiesche
Department of Mechanical Engineering
Muenster University of Applied Sciences
Steinfurt, Germany

Christian Helcig
Department of Mechanical Engineering
Muenster University of Applied Sciences
Steinfurt, Germany

ISSN 2191-530X   ISSN 2191-5318 (electronic)
SpringerBriefs in Applied Sciences and Technology
ISBN 978-3-319-20166-5   ISBN 978-3-319-20167-2 (eBook)
DOI 10.1007/978-3-319-20167-2

Library of Congress Control Number: 2015943320

Springer Cham Heidelberg New York Dordrecht London

Springer International Publishing AG Switzerland is part of Springer Science+Business Media (www.springer.com)

Preface

This book is devoted to the flow and convective heat transfer phenomena that occur when a rotating disk is subjected to an outer stream of air. Special attention is given to the effects due to the angle of attack for inclined rotating disks. Such a configuration represents a direct generalization of the fairly classic problems for free rotating disks without any outer forced flows or for flow impingement onto an orthogonal rotating disk. These both axis-symmetric configurations have been widely discussed in the scientific literature, but little research has been done on inclined rotating disks passed by an outer forced flow. This topic was therefore the subject of a research project funded by the *Deutsche Forschungsgemeinschaft DFG*. The research provides insight into the complex configuration characterized by interactions between rotating surfaces and boundary layers, flow separation and transition phenomena, and into the establishment of distinct flow and heat transfer regimes.

The main experimental and theoretical work of the project was finished in 2014. The research showed that flow and heat transfer from inclined rotating disks subjected to an outer stream of air delivers excellent examples for studying flow transition and bifurcation phenomena in more detail. Therefore, the results might be interesting not only for the scientific and engineering community directly faced with rotating disk problems but also for a large number of fluid dynamics researchers as well. The present contribution to the SpringerBriefs in Applied Sciences and Technology makes the results available to a wider audience.

Steinfurt, Germany

Stefan aus der Wiesche
Christian Helcig

Acknowledgments

The authors gratefully acknowledge the *Deutsche Forschungsgemeinschaft DFG* for its strong financial support to the present work. This support has made it possible to systematically collect and evaluate data on an inclined rotating disk subjected to a stream of air.

The research involved many individuals from the laboratory team at Muenster University of Applied Sciences. Their efforts and contributions are highly appreciated. In particular, thanks go to the former undergraduate students Claus-Martin Trinkl, Ufuk Bardas, Alexander Weyck, Lasse Nagel, and Christian Teigeler for their contributions.

A multi-year research study cannot be successfully performed without the feedback and comments of distinguished experts. Among the many scientists and workers on the field, the authors would like to deeply acknowledge the fruitful discussions with Igor V. Shevchuk. His comments and suggestions were extremely valuable throughout the study. Furthermore, the authors would also like to thank Professor Weigand, University of Stuttgart, for his encouraging interest and valuable suggestions. A great opportunity for presentation and discussion of results were given by the conferences organized by the *American Society of Mechanical Engineers ASME*. Particularly, the contributions and the encouraging feedback from the *ASME Fluids Engineering Division* were a great support.

The editorial assistance of the staff at Springer and the interests of the editor of the series, Professor Kulacki, were also gratefully appreciated.

Finally, one of the authors, Stefan aus der Wiesche, would like to express his special thanks to his wife Julia for her support and patience.

Contents

Nomenclature

a	Parameter for ellipsoid
a	Potential flow parameter
a_T	Thermal diffusivity
A	Variable for functional
A_i	i-th area sector
A_i	Tensor (Taylor series for velocity)
A_{ij}	Tensor (Taylor series for velocity)
A_{ijk}	Tensor (Taylor series for velocity)
b	Parameter for ellipsoid
B	Angular velocity ratio
B	Variable
c	Parameter for ellipsoid
c_p	Isobaric specific heat
C	(Correlation) constant
C	Variable
C_{cr}	Correlation constant (Landau model)
C_D	Drag coefficient
C_i	Correlation constant ($i = 1, 2, 3$)
C_{nc}	Correlation constant for natural convection heat transfer
C_S	Smagorinsky constant
d	Thickness
d	Diameter
f	Function
F	Function
F	Self-similar function (radial component)
F_D	Drag force
g	Acceleration due to gravity
g_i	i-th weighting factor
G	Function

G	Self-similar function (azimuthal component)
$G_{\Delta x}$	Filter function (LES)
Gr	Grashof number
h	Heat transfer coefficient
h_m	Mean heat transfer coefficient
h_x	Local heat transfer coefficient based on coordinate x
H	Self-similar function (axial component)
H_1	Self-similar function (axial component)
H_2	Self-similar function (axial component)
K	Correlation constant
K_m	Correlation constant (mean heat transfer)
m	Correlation exponent
m_{cr}	Correlation exponent (Landau model)
m_i	Correlation exponent ($i = 1, 2, 3$)
Ma	Mach number
n_R	Correlation exponent
n^*	Exponent for temperature distribution function
N	Velocity ratio (potential flow)
N	Total number
Nu	Nusselt number
Nu_m	Mean Nusselt number
$Nu_{m,a}$	Actual mean Nusselt number
$Nu_{m,nc}$	Mean Nusselt number for natural convection
Nu_{2R}	Nusselt number based on disk diameter $2R$
p	Pressure
p_s	Stagnation pressure
$\overline{p}^*$	Macro pressure (LES)
P	Self-similar function (pressure)
Pr	Prandtl number
Pr_t	Turbulent Prandtl number
q_i	Subgrid flux component in i-direction (LES)
$\dot{q}_w$	Heat flux
r	Radial coordinate
r_i	ith radial coordinate sector
R	Radius
R_{cr}	Critical ratio between Reynolds numbers (Landau model)
R^*	Radius for primary rotational flow domain
Re	Reynolds number
Re_a	Reynolds number based on potential flow parameter a
$Re_{a,2R}$	Reynolds number based on potential flow parameter a and disk diameter $2R$
$Re_{d,cr}$	Critical Reynolds number based on sphere diameter d
Re_e	Combined Reynolds number
Re_u	Inflow or translational Reynolds number
Re_δ	Reynolds number based on boundary layer thickness

Re_ω	Rotational Reynolds number
$Re_{\omega,r}$	Local rotational Reynolds number based on radial coordinate r
$Re_{\omega,2R}$	Rotational Reynolds number based on disk diameter $2R$
S	Strain rate tensor
S_{ij}	Strain rate tensor component
Sc	Schmidt number
Sh	Sherwood number
t	Time
T	Temperature
T_{ij}	Subgrid tensor (LES)
TF	Turbulence factor
Tu	Turbulence intensity
u	Velocity (component)
u'	Fluctuating part of velocity
u_i	Velocity component in i-direction
u_{jet}	Jet velocity
x	Coordinate
x_i	Coordinate in i-direction
$\mathbf{x}$	Coordinate vector
y	Coordinate
$\mathbf{y}$	Coordinate vector
z	Axial coordinate

Greek Symbols

α	Diffusivity coefficient
α_t	Eddy diffusivity coefficient (LES)
α_0	Parameter for velocity potential
β	Angle of attack, incidence
β_T	Thermal expansion coefficient
Γ	Circulation
δ_1	Boundary layer thickness
ζ	Self-similar variable
ζ	Normal coordinate
θ	Passive variable (e.g., temperature)
θ_s	Alignment angle
Θ	Normalized temperature
λ	Thermal conductivity
λ	Parameter for ellipsoid
λ_{pert}	Perturbation wavelength
Λ	Order parameter
μ	Dynamic viscosity
ν	Kinematic viscosity
ν_t	Eddy viscosity

ρ	Density
ϕ	Velocity potential
ψ	Functional
Ψ	Control parameter
ω	Angular velocity (disk)
Ω	Angular velocity (flow)

Subscripts

av	Average
cp	Constant properties
f	Film
i	i-Direction ($i = 1, 2, 3$)
j	j-Direction ($j = 1, 2, 3$)
k	k-Direction ($k = 1, 2, 3$)
l	Laminar
m	Mean
M	Maximum
r	Radial
ref	Reference
t	Turbulent
tr	Transition
tr,ω	Transition with regard to ω
w	Wall
z	Axial
φ	Azimuthal
0	Reference, nominal
∞	Infinity, ambient, bulk

Superscript

+	Normalized

Mathematical Symbols

$\delta\psi$	Variation of ψ
ΔT	Temperature difference
Δu_i	Laplace operator for velocity component u_i
$\overline{f}$	Filtering of function f (LES)
f'	Fluctuating part of variable f (LES)
F'	Derivative of self-similar function F

Chapter 1
Introduction

Configurations based on rotating disks are widely used in engineering applications, and convective heat transfer from a rotating disk has been studied extensively in the scientific literature for several decades. In addition to a large number of technical papers, several monographs [1–4] about this topic are also available. Today, research dedicated to flow and heat transfer phenomena in rotating disk systems is still being performed, as demonstrated by the increasing number of corresponding scientific and technical conference contributions and research articles.

In certain respects, the flow in the vicinity of a rotating surface resembles the flow over a stationary surface if the frame of reference of the observer is moving with the rotating surface. However, in many cases, substantial additional flow phenomena arise due to the action of centrifugal or Coriolis forces or due to the nature of a three-dimensional boundary layer flow. Then, the resulting flow characteristics can be very complex, and according to Kreith [5], "almost every rotating system reveals novel and unexpected flow characteristics when subjected to a complete stability analysis or when studied experimentally over a sufficiently wide range of all the variables which affect the hydrodynamic phenomena." Since in rotating systems convective heat transfer and flow are intimately related, they are also complex and present interesting scientific and practical phenomena. It is therefore not surprising that much research has been dedicated to flow and heat transfer in rotating systems.

In rotating disk systems, the two classes of (1) enclosed rotating disks or rotating cavities and (2) the basic configuration of a free rotating disk can be identified. The first class of rotating disk systems occurs frequently in gas turbine engineering. With the advent of air-cooled gas turbines, the investigation of flow and heat transfer phenomena due to rotor–stator interactions became mandatory for the gas turbine industry. The latter case of a free rotating disk is from an academic or fundamental point of view of major importance, but it represents as well an obvious

© The Author(s) 2016
S. aus der Wiesche, C. Helcig, *Convective Heat Transfer From Rotating Disks Subjected To Streams Of Air*, SpringerBriefs in Applied Sciences and Technology, DOI 10.1007/978-3-319-20167-2_1

starting point for analyzing numerous technical applications in turbo machinery, computer hard disks, CVD reactors, train wheel, or disk brake design.

The majority of the investigations consider enclosed rotating disk or free rotating disks with or without an outer forced flow perpendicular to the disk plane. Much less attention has been paid to configurations based on a rotating disk subjected to an outer flow parallel to the plane of rotation, and the heat transfer from an inclined rotating disk (i.e., with an angle of attack in respect to the uniform stream) has not been covered by review articles or monographs so far. This book considers a series of flow and heat transfer phenomena that occur when an inclined free rotating disk is subjected to an outer forced stream of air, and special attention is given to the effect of the angle of attack.

The possible flow configurations for a free disk subjected to a uniform flow are illustrated in Fig. 1.1. The axisymmetric flow field for an orthogonal flow is shown in Fig. 1.1a. A vanishing outer flow results in the limit case of a free rotating disk in a resting fluid. This configuration has been intensively considered in the scientific literature since the pioneering work [6] of von Karman in 1921. Whereas the forced flow perpendicular to the disk leads to an axisymmetric stagnation flow, see Fig. 1.1a, the combination of rotation and a parallel flow, as in Fig. 1.1b, leads to a non-axisymmetric, fully three-dimensional flow. The parallel flow over a disk with finite thickness is characterized by flow separation at the leading disk edge followed by reattachment of a turbulent boundary as illustrated in Fig. 1.1b.

In 1970, Dennis et al. [7] studied for the first time heat transfer coefficients for a large range of rotational and crossflow Reynolds numbers by placing an electrically heated disk in the parallel air stream of a wind tunnel. They reported "that the freestream turbulence in the tunnel must have been high", perhaps yielding systematically higher heat transfer rates. In 1974, Booth and de Vere [8] conducted a comprehensive set of measurements of the radial variation of the heat transfer coefficients for the same configuration but obviously without knowledge of the prior work [7]. The authors concluded that "in any situation the level of heat transfer coefficients is determined in the main by the speed of the transverse air flow." Such a statement at least partially contradicts the findings of Dennis et al. [7].

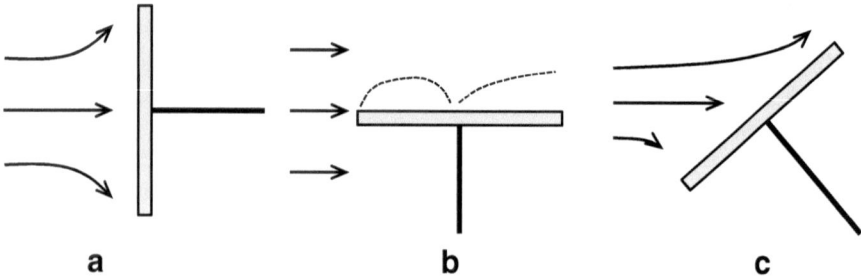

Fig. 1.1 Schematics of the flow fields caused by the outer forced flow for a stationary disk. Axisymmetric configuration (**a**) parallel disk with flow separation (**b**) inclined disk (**c**)

In 2005, He et al. [9] employed the naphthalene sublimation technique to obtain the local Sherwood number of a rotating disk for a limited range of a local Reynolds number, but unfortunately they did not distinguish between the rotational and the crossflow Reynolds numbers. Results of an extensive large-eddy-simulation (LES) study were published by one of the authors [10], but his numerical data indicated heat transfer coefficients much lower than those measured experimentally by Dennis et al. [7]. Since the deviations between the available data were substantial, an independent re-investigation has been performed, and corresponding experimental data [11] were published in 2011.

The general case of an inclined rotating disk subjected to an outer flow as illustrated in Fig. 1.1c has rarely been considered in the literature. Some experimental data were presented in [11], but a systematic study of the effect of the angle of attack on the flow and convective heat transfer is still lacking. The number of articles dealing with flow and heat transfer over rotating disks has rapidly increased over the last decades, but this canonical configuration remained surprisingly unexplored. It was thus the subject of a research study funded by the *Deutsche Forschungsgemeinschaft DFG*, and the major results of this systematic investigation are presented in this book. Obviously, an inclined disk represents the natural connection between the perpendicular and the parallel disk, and it therefore connects two configurations with substantially different symmetry properties. This aspect makes it directly interesting from a theoretical point of view. From a practical point of view, the affect the incidence angle has on the flow and heat transfer is important in many applications.

The flow and convective heat transfer from an inclined disk is governed in great parts by the phenomenon of flow separation and by the interaction of the boundary layer flow with the rotating disk surface. While separation and the transition to turbulence are well understood in two-dimensional boundary layer flows, there is still no complete theory available for three-dimensional flow separation. However, in the past, critical point and bifurcation theory have provided a good strategy for a qualitative discussion of phenomena occurring due to three-dimensional flow separation, and in the following we show that this approach is also valuable for the present configuration of an inclined rotating disk subjected to a uniform forced flow.

All theoretical considerations are generally based upon measurements and experimental research. In the present case, only air at atmospheric conditions was considered as the fluid, and experiments were carried out in test sections of low-speed wind tunnels. This approach is therefore discussed in some detail. To an increasing extent, computational fluid dynamics (CFD) methods are now being used to simulate flow and heat transfer in rotating disk systems. Although the basic configuration of a free rotating disk in a uniform stream of air seems to be fairly simple, modeling and simulation of flow separation and the corresponding effects are still challenging for CFD. Here, LES methods offer great potential. This promising, modern approach will be discussed briefly in this contribution, too.

References

1. Dorfman LA (1963) Hydrodynamic resistance and the heat loss of rotating solids. Oliver & Boyd, Edinburgh
2. Owen JM, Rogers RH (1989) Flow and heat transfer in rotating disc systems, vol 1, Rotor-stator systems. Research Studies, Taunton
3. Owen JM, Rogers RH (1989) Flow and heat transfer in rotating disc systems, vol 2, Rotating cavities. Research Studies, Taunton
4. Shevchuk IV (2009) Convective heat and mass transfer in rotating disk systems. Springer, Berlin
5. Kreith F (1968) Convection heat transfer in rotating systems. Adv Heat Tran 5:129–251
6. von Karman T (1921) Über laminare und turbulente Reibung. ZAMM 1(4):233–252
7. Dennis RW, Newstead C, Ede AJ (1970) The heat transfer from a rotating disc in an air crossflow. In: Proceedings of 4th International Heat Transfer Conference, Versailles, 1970 (paper FC 7.1)
8. Booth GL, de Vere APC (1974) Convective heat transfer from a rotating disc in a transverse air stream. In: Proceedings of 5th International Heat Transfer Conference, Tokyo, 1974 (paper FC1.7)
9. He Y, Ma L, Huang S (2005) Convection heat and mass transfer from a disk. Heat Mass Transf 41:766–772
10. aus der Wiesche S (2007) Heat transfer from a rotating disk in a parallel air crossflow. Int J Therm Sci 46:745–754
11. Trinkl CM, Bardas U, Weyck A, aus der Wiesche S (2011) Experimental study of the convective heat transfer from a rotating disc subjected to forced air streams. Int J Therm Sci 50:73–80

Chapter 2
Basic Principles

This chapter briefly reviews some basic principles that are useful when studying flow and heat transfer in rotating disk systems. Further details about these principles can be found in the literature, and it is therefore not necessary to present all of their implications here. The flow over an inclined disk with finite thickness is in general three-dimensional and characterized by flow separation. The transitions between different flow and heat transfer regimes can be described in terms of the critical point theory proposed for the first time in the early 1950s. This approach has been developed further over the last decades, and its connections to bifurcation theory became apparent. Its great potential for analyzing rotating disk systems has been recently realized, and its basic principles are given here.

2.1 Governing Equations

In this book, the air is considered to be an incompressible, Newtonian fluid with practically constant material properties. The assumption of an incompressible fluid is justified for low speed levels corresponding to Mach numbers below $Ma < 0.2$ [1], giving, for air at atmospheric conditions, a maximum velocity limit between 60 and 80 m/s. Although in air there is a temperature dependency of the material properties such as viscosity or thermal conductivity, these effects can be neglected at small or moderate temperature differences within the system. For disks with high thermal loads, this rather simple assumption should be replaced by a more appropriate treatment taking the temperature dependency of the quantities into account [2]. For fluid flows over disks rotating with sufficiently high angular velocities ω or subjected to strong streams, contributions due to natural convection or effects of gravitational forces on momentum transfer can be neglected.

Under the above assumptions, the governing flow equations are given by the Navier Stokes equations that call in the usual Cartesian tensor notation

© The Author(s) 2016
S. aus der Wiesche, C. Helcig, *Convective Heat Transfer From Rotating Disks Subjected To Streams Of Air*, SpringerBriefs in Applied Sciences and Technology, DOI 10.1007/978-3-319-20167-2_2

$$\frac{\partial u_i}{\partial t} + u_j \frac{\partial u_i}{\partial x_j} = -\frac{1}{\rho}\frac{\partial p}{\partial x_i} + \frac{\mu}{\rho}\frac{\partial^2 u_i}{\partial x_j^2} \qquad (2.1)$$

and the continuity equation

$$\frac{\partial u_i}{\partial x_i} = 0. \qquad (2.2)$$

Since heat transfer is involved, the energy equation has to be considered. It can be written under the above assumptions as a partial differential equation

$$\frac{\partial T}{\partial t} + u_i \frac{\partial T}{\partial x_i} = a_T \frac{\partial^2 T}{\partial x_i^2}. \qquad (2.3)$$

for the temperature field T in the flow domain. In addition to the governing equations (2.1)–(2.3), appropriate boundary conditions have to be given. For rotating solids, the no-slip condition is usually applied to the velocity field. With regard to heat transfer, a prescribed wall heat flux $\dot{q}_w$ or a prescribed wall temperature distribution T_w are frequently used boundary conditions. Since the classic investigation by Dorfman, the assumption of a temperature distribution in accordance to a power law

$$T_w = T_0 + C r^{n*}. \qquad (2.4)$$

with constant parameters T_0, C, and $n*$ is common for analytical treatments [3]. The isothermal case with constant wall temperature T_w corresponds to $n* = 0$.

For axisymmetric configurations such as free rotating disks in a resting fluid or flow impingement onto orthogonal disks, a cylindrical coordinate system can be usefully applied. This configuration with its main parameters is shown in Fig. 2.1.

In Fig. 2.1, the rotation axis of the disk is chosen as z-coordinate axis, while the point $z = 0$ is located at the disk surface. In this case, the governing equations for stationary flow are explicitly given by the following set of partial differential equations:

$$u_r \frac{\partial u_r}{\partial r} + u_z \frac{\partial u_r}{\partial z} - \frac{u_\varphi^2}{r} = -\frac{1}{\rho}\frac{\partial p}{\partial r} + \frac{\mu}{\rho}\left(\Delta u_r - \frac{u_r}{r^2}\right), \qquad (2.5)$$

Fig. 2.1 Axisymmetric configuration and its main parameters (flow and temperature profile over a rotating disk in still air)

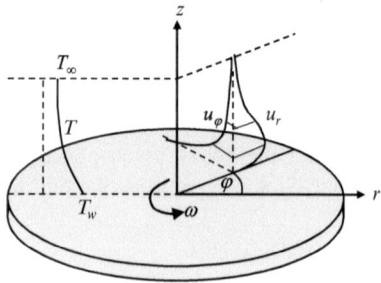

$$u_r \frac{\partial u_\varphi}{\partial r} + u_z \frac{\partial u_\varphi}{\partial z} + \frac{u_r u_\varphi}{r} = \frac{\mu}{\rho}\left(\Delta u_\varphi - \frac{u_\varphi}{r^2}\right), \tag{2.6}$$

$$u_r \frac{\partial u_z}{\partial r} + u_z \frac{\partial u_z}{\partial z} = -\frac{1}{\rho}\frac{\partial p}{\partial z} + \frac{\mu}{\rho}\Delta u_z, \tag{2.7}$$

$$\frac{\partial u_r}{\partial r} + \frac{u_r}{r} + \frac{\partial u_z}{\partial z} = 0 \tag{2.8}$$

with the operator

$$\Delta = \frac{\partial^2}{\partial r^2} + \frac{1}{r}\frac{\partial}{\partial r} + \frac{\partial^2}{\partial z^2}. \tag{2.9}$$

In a rotating coordinate system (i.e., a frame fixed with the disk), the governing equations include additional Coriolis force terms. For further details about formulating the corresponding Navier–Stokes equations in moving or rotating frames, the reader is referred to [3].

2.2 Boundary-Layer Approach

In fluid mechanics, the boundary-layer theory is of fundamental importance, and many problems can only be treated successfully within its framework [4]. In the case of a rotating disk in still air, the usual boundary-layer assumptions are [3, 4]:

1. The velocity component u_z is by an order of magnitude lower than the other components.
2. The change of velocity, pressure, and temperature in the normal direction is much larger than in the radial direction.
3. The static pressure is constant in the normal direction.

Under these assumptions, it is possible to replace the Navier–Stokes equations (2.6) and (2.7) by the following set of boundary-layer equations

$$u_r \frac{\partial u_r}{\partial r} + u_z \frac{\partial u_r}{\partial z} - \frac{u_\varphi^2}{r} = -\frac{1}{\rho}\frac{\partial p}{\partial r} + \frac{\mu}{\rho}\frac{\partial^2 u_r}{\partial z^2}, \tag{2.10}$$

$$u_r \frac{\partial u_\varphi}{\partial r} + u_z \frac{\partial u_\varphi}{\partial z} + \frac{u_r u_\varphi}{r} = \frac{\mu}{\rho}\frac{\partial^2 u_\varphi}{\partial z^2}, \tag{2.11}$$

$$\frac{\partial p}{\partial z} = 0. \tag{2.12}$$

The continuity equation (2.8) remains invariable. The equation for a stationary thermal boundary layer is

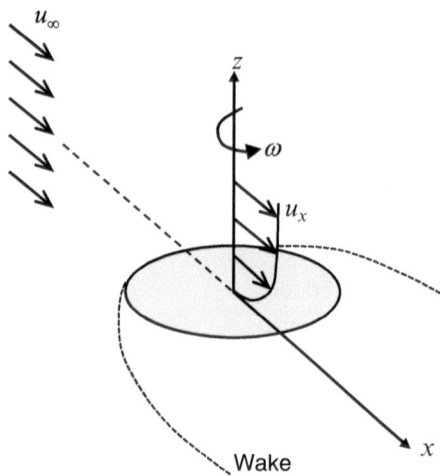

Fig. 2.2 Flow over a parallel rotating disk

$$u_r \frac{\partial T}{\partial r} + u_z \frac{\partial T}{\partial z} = a_T \frac{\partial^2 T}{\partial z^2}. \tag{2.13}$$

The above set of equations is closed by the equation

$$\frac{1}{2}\frac{du_{r,\infty}^2}{dr} - \frac{u_{\varphi,\infty}^2}{r} = -\frac{1}{\rho}\frac{dp_\infty}{dr} \tag{2.14}$$

that connects the flow variables in the potential flow domain.

For a thin disk subjected to a parallel stream of air, the boundary-layer approach has to be applied to the flow arrangement illustrated in Fig. 2.2, which differs substantially from the situation shown in Fig. 2.1 and is described by the set of equations (2.10)–(2.14).

For parallel or inclined disks with finite thickness, wakes and three-dimensional flow separation become essential. While the phenomenon of separation in two-dimensional flow is fairly well understood, the situation in three dimensions is much more complicated and far from being clear. Furthermore, a moving solid boundary wall (i.e., the rotating disk surface) changes the separation condition. Then, Prandtl's well-known two-dimensional separation condition

$$\left.\frac{\partial u}{\partial z}\right|_w = 0 \tag{2.15}$$

at the wall has to be replaced by the so-called MRS-criterion [5–7]

$$\frac{\partial u}{\partial z} = 0 \text{ at regions with vanishing velocity } u = 0, \tag{2.16}$$

satisfied typically within the flow domain. The separated flow is usually unsteady, and the laminar boundary-layer theory does not provide a fully efficient approach for calculating the entire flow field. For these types of flows and phenomena, the critical point and bifurcation theory offers a valuable approach that will be discussed in more detail in following sections of this chapter.

The integral method is a classic and powerful approach to solving stationary axisymmetric boundary-layer problems arising from rotating disks. Commonly, the first step in this method is to replace the differential equations with integral boundary-layer equations (which are in fact integrodifferential equations). In a second step, the integral boundary-layer equations are solved using appropriate models for the involved velocity and wall shear-stress components. Models for temperature (or enthalpy thickness) profiles and wall heat flux expressions are required for solving the thermal boundary-layer problem. It is possible to account for laminar and turbulent flow. The integral method is capable of solving a large variety of flow and heat transfer problems in rotating disk systems [3, 8, 9], but applying this method to fully three-dimensional flow including flow separation and reattachment of turbulent boundary layers is difficult or nearly impossible.

2.3   Self-Similar Solutions

For laminar flows over a single rotating disk, exact solutions of the Navier–Stokes equations can be obtained by means of the concept of self-similar solutions. Since the pioneering work [10] by von Karman in 1921, this class of exact solutions has been also termed von Karman swirling flows, and a review of its mathematical properties has been given by Zandbergen and Dijkstra [11]. For a free rotating disk subjected to a perpendicular stream of air, the following self-similar variables are introduced:

$$u_r = (a + \omega) r F(\zeta), \tag{2.17}$$

$$u_z = \sqrt{(a + \omega) \nu}\, H(\zeta), \tag{2.18}$$

$$u_\varphi = (a + \omega) r\, G(\zeta), \tag{2.19}$$

$$p = -\rho \nu \omega P(\zeta) \tag{2.20}$$

with the new coordinate

$$\zeta = z \sqrt{\frac{a + \omega}{\nu}}. \tag{2.21}$$

The functions F, G, H, and P obey a set of *ordinary* differential equations

$$F^2 - G^2 + F'H = \frac{N^2 - B^2}{(1 + N)^2} + F'', \tag{2.22}$$

$$2FG + G'H = G'',\tag{2.23}$$

$$HH' = P' + H'',\tag{2.24}$$

$$2F + H' = 0,\tag{2.25}$$

derived from the boundary-layer equations. The potential flow variable a is connected with the assumed flow

$$\zeta \to \infty: \quad u_{r,\infty} = ar, \quad u_{z,\infty} = -2az, \quad u_{\varphi,\infty} = \Omega r \tag{2.26}$$

at infinity and has the dimension of 1/s. The ratio between the disk angular velocity ω and the outer flow rotation Ω is denoted by parameter $B = \Omega/\omega$ and is assumed to be constant. The non-dimensional radial velocity in the potential flow outside the boundary layer is given by $N = u_{r,\infty}/\omega \ r = a/\omega$. At the disk surface, the boundary conditions are

$$\zeta = 0: \quad F = H = 0, \quad G = 1. \tag{2.27}$$

In the past, the above set of equations has been solved by means of expansions in power or exponential series and by using the shoot-method. Today, modern mathematical software tools enable user-friendly handling of such sets of ordinary differential equations. The exact solutions provide a reliable database for validation studies of CFD methods. Using self-similar solutions also provides a good opportunity for obtaining approximate analytical solutions for other configurations, but it should be remarked that currently this approach has been mainly replaced by numerical methods (CFD) in engineering applications.

2.4 Dimensional Analysis and Correlations

The general behavior of all fluid mechanical systems can be best understood through dimensional analysis. This is the formal procedure whereby the group of involved variables representing the flow and heat transfer configuration is reduced to a smaller number of dimensionless groups. When the number of independent variables is not too great, dimensional analysis enables a clear and transparent picture. Several methods for constructing non-dimensional groups have been described [12]. In the case of flow and convective heat transfer from an inclined rotating disk as illustrated in Fig. 2.3, the following groups are useful based on the main parameters of the considered problem.

The mean heat transfer coefficient h_m is used for the definition of a mean or average Nusselt number with the disk radius R

Fig. 2.3 Main parameters
for the flow and heat
transfer from an inclined
rotating disk

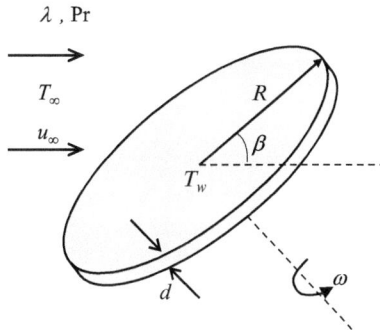

$$Nu_m = \frac{h_m R}{\lambda} \tag{2.28}$$

with

$$h_m = \frac{\dot{q}_{w,m}}{T_{w,m} - T_\infty} \quad \text{and} \quad \dot{q}_{w,m} = \frac{1}{\pi R^2} \int_0^{2\pi} \int_0^R \dot{q}_w r \, dr \, d\varphi. \tag{2.29}$$

Instead of a mean Nusselt number Nu_m, the use of a local Nusselt number

$$Nu = \frac{\dot{q}_w r}{\lambda (T_w - T_\infty)} \tag{2.30}$$

is also common, employing local values at the radial coordinate r.

The flow behavior is described by an inflow or translational Reynolds number

$$Re_u = \frac{u_\infty R}{v} \tag{2.31}$$

based on the velocity u_∞ of the outer uniform forced flow and a rotational Reynolds
number

$$Re_\omega = \frac{\omega R^2}{v}. \tag{2.32}$$

Sometimes, a local rotational Reynolds number

$$Re_{\omega,r} = \frac{\omega r^2}{v} \tag{2.33}$$

based on the radial coordinate r is also used (in many cases in connection with a
local Nusselt number (2.30)).

The thermal fluid properties are generally described by the Prandtl number

$$Pr = \frac{\nu}{a_T}, \tag{2.34}$$

but in this book only dry air at atmospheric conditions with $Pr = 0.71$ is considered. Since incompressible flow is assumed, the Mach number is equal to zero, $Ma = 0$. Body forces or natural convection effects are also neglected in the present analytical treatment, and hence no Grashof or Rayleigh numbers are required, but in light of the actual experiments and their results, the role of natural convection has to be carefully considered (see Chap. 3).

The geometric arrangement is described by the angle of attack (or incidence) β. This variable is of major importance because it governs the symmetry. The case $\beta = 90°$ corresponds to the axisymmetric stagnation flow, whereas the three-dimensional flow over a parallel disk would be represented by $\beta = 0°$.

Of further influence is the thickness ratio d/R of the disk because, together with the angle of attack β, this parameter governs the location of a stagnation point and hence the flow regime over the disk.

The resulting flow regime also depends on the turbulence level of the inflow and the roughness of the disk surface because under certain circumstances these parameters govern the transition from a laminar boundary-layer flow to a turbulent boundary layer.

In summary, the mean Nusselt number for a thin rotating disk should be expressed at least as the function

$$Nu_m = Nu_m(Re_u, Re_\omega, \beta) \quad Pr = 0.71, \quad Ma = 0 \tag{2.35}$$

of the involved three major parameters. However, we demonstrate later that other minor parameters such as d/R or the inflow turbulence level Tu are also, under certain circumstances, of some importance for heat transfer correlations.

2.5 Topology of Three-Dimensional Separated Flows

Three-dimensional separated flow represents a domain of great practical interest that is still beyond the reach of definitive theoretical analysis or numerical computations. At present, the understanding of three-dimensional flow separation rests on observations drawn from experiments. However since the pioneering works in the early 1950s by Legendre [13–15], the topology of such flows can also be described in terms of a critical point theory that is particularly useful for classifying and interpreting transitions between different flow and heat transfer regimes of rotating disk systems.

Describing and understanding three-dimensional flows present challenges because of the use of inappropriate terms linked to two-dimensional flows. In the case of rotating surfaces, Prandtl's classic two-dimensional description has only limited value, but as pointed out by Legendre [16] in 1956, the pattern of stream-lines immediately adjacent to the surface might be considered as trajectories having properties consistent with those of a continuous vector field. On the basis of his postulate, it follows that there is a strong connection to the theory of autonomous differential equations and bifurcation theory. This can be demonstrated by consid-ering the Navier–Stokes equations (2.1) and the continuity equation (2.2) for the flow field. Assuming that the solutions of that set of equations are regular at all points, an arbitrary point O in the flow field can be chosen and a Taylor series

$$u_i = A_i + A_{ij}x_j + A_{ijk}x_jx_k + \ldots = \dot{x}_i \qquad (2.36)$$

can be used to expand the velocity u_i in terms of the coordinates x_j with the origin for x_j located at O. The coefficients A_i, A_{ij}, ... are functions of time if the flow is unsteady, and they are symmetric tensors in all indices except the first. Except at singular points, it follows from (2.36) that at a point O very close to a rigid surface, the velocity components must grow linearly with a coordinate ζ directed out of the surface normal to the general curvilinear coordinates being orthogonal axes in the surface. The velocity components of a streamline near the surface obey then a set of autonomous ordinary differential equations [13]. If O is located at a critical point, then the streamline slope is indeterminate, which means that the lowest order coefficients A_i are equal to zero. There are two types of critical points [14], namely (1) no-slip critical points such as a separation point on a no-slip boundary ($A_{ij} = 0$), and (2) free-slip critical points occurring within the fluid away from the surface (A_{ij} finite). No-slip critical points are classifiable into nodes and saddle points, and nodes may be further subdivided into two subclasses (nodal points and foci of separation or attachment) as discussed by Lighthill [17] in more detail. Mathematically, they are governed by the properties of the tensors A_{ijk}.

In the case of inclined rotating disks subjected to a stream of air, parameters such as the angle of attack or the Reynolds numbers determine the three-dimensional flow separation. In line with the terminology of Andronov et al. [18], the pattern of skin friction lines on the body constitutes the phase portrait of the surface shear-stress vector. Two phase portraits have the same topological structure if a mapping from one to the other phase portrait preserves the paths. A topological property is any characteristic of the phase portrait that remains invariant under all path-preserving mappings. The number and types of critical points are examples of such topological properties. The set of all topological properties of the phase portrait describes the topological structure.

The most important issue for the present purpose is the *structural stability* of phase portraits relative to a *control parameter* Ψ. The control parameter Ψ can be given directly by a relevant variable that governs the flow regime (e.g., $\Psi = \beta$) or it can be given by suitable non-dimensional groups (e.g., $\Psi = Re_\omega/Re_u$). A phase portrait is structurally stable at a given value of the control parameter Ψ if the phase portrait resulting from an infinitesimal change in the parameter has the same topological structure. It is useful to distinguish between structural stability and *asymptotic stability* of the flow. A mean flow is called asymptotically stable if small perturbations from it at a fixed value of the control parameter γ decay to zero as time $t \to \infty$. Furthermore, it is useful to distinguish between *local* and *global* properties of the instabilities. Instability is called global if it permanently alters the topological structure of either the external three-dimensional velocity field or the phase portrait of the surface-stress field. Instability is local if it does not result in a change of the topological structure of vector fields. A structural instability is always global while an asymptotic instability implies non-uniqueness in the solutions of the governing flow equations.

Asymptotic instability of the external flow leads to the notions of bifurcation, symmetry breaking, and dissipative structures that have been applied to rotating disk systems [19]. The fluid motions evolve according to the time-dependent equations of the general form

$$\frac{\partial u_i}{\partial t} = G(u_j, \Psi) \tag{2.37}$$

that is governed by the control parameter Ψ. The steady mean flow is represented by

$$0 = G(u_j, \Psi). \tag{2.38}$$

A mean flow $u_{0,i}$ is an asymptotically stable flow if small perturbations from it decay with increasing time. When the control parameter Ψ is varied, one mean flow $u_{0,i}$ may persist in the mathematical sense that it remains a valid solution of (2.38) but becomes unstable to small perturbations as the control parameter Ψ crosses a critical or transitional value Ψ_{tr}. At such a transition point $\Psi = \Psi_{tr}$, a new mean flow may bifurcate from the known flow $u_{0,i}$. It is assumed that the flow and heat transfer behavior can be sufficiently described by an *order parameter* or amplitude Λ. Such an order parameter Λ might be given by a mean Nusselt number related to a reference value, or by another quantity that characterizes the bifurcation flow alone (e.g., the amplitude of an oscillating velocity component [20]). The behavior is well illustrated by means of the bifurcation diagram shown in Fig. 2.4. Stable flows are indicated by solid lines in Fig. 2.4 and unstable flows by dashed lines. Over the range $\Psi < \Psi_{tr}$ where the flow is stable, the order parameter Λ remains constant. The flow becomes unstable for all values $\Psi > \Psi_{tr}$, as the dashed line on the abscissa in Fig. 2.4 indicates. A new mean flow bifurcates from $\Psi = \Psi_{tr}$ either *supercritically* or *subcritically*. At a supercritical bifurcation, the new flow that replaces the unstable one differs infinitesimally from it. As Ψ increases beyond Ψ_{tr}, the bifurcation flow departs significantly from the unstable flow. The bifurcation breaks the symmetry of the known flow, adopting a form of lesser symmetry in which dissipative structures arise to absorb just the amount of excess energy that the more symmetrical flow is no longer able to absorb.

At a subcritical bifurcation, there are no adjacent new flows that differ only infinitesimally from the unstable flow. Here, a finite jump in the development of the order parameter Λ occurs at $\Psi = \Psi_{tr}$ for increasing Ψ. This finite jump

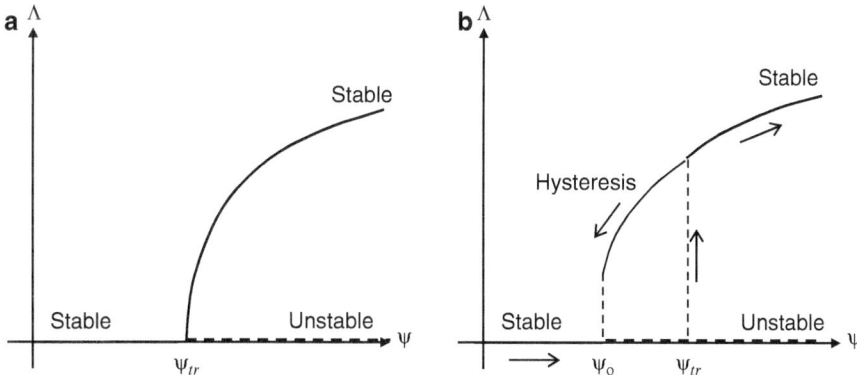

Fig. 2.4 Examples of a supercritical bifurcation (**a**) and a subcritical bifurcation (**b**)

represents a radical change in the topological structure of the external flow and perhaps in the phase portrait of the surface-stress field as well. When Ψ is then decreased, the new flow may persist up to a value $\Psi_o < \Psi_{tr}$. For $\Psi < \Psi_o$ the stable flow is recovered. Hysteresis effects occur in a subcritical bifurcation, and a strong analogy to the phenomena of meta-stable states in thermodynamics becomes obvious. In line with the analogy of phase transitions in thermodynamics, a supercritical bifurcation might be also called Landau-transition, whereas a subcritical bifurcation could be described by the Landau-de Gennes model [19], but special attention should be paid to the values of the critical exponents of the bifurcation. The Landau model based on a mean-field approximation, and hence the exponent in the increase of the order parameter Λ as a function of the control parameter Ψ has the special value 1/2 [20].

The classic example of a supercritical bifurcation is given by the flow over a round-nosed body of revolution at the angle of attack [13]. In rotating disk systems, both types of bifurcation occur, and some of them will be outlined in detail in the present book. In the case of a parallel free rotating disk subjected to a stream of air, a supercritical bifurcation model was first proposed by aus der Wiesche [21] in 2002 for the heat transfer augmentation due to rotation. This model was supported by large-eddy-simulations [20, 22]. Latour et al. [23] also demonstrated the value of such a model for convective heat transfer from a rotating disk mounted on a shaft and subjected to an air crossflow. For an inclined resting disk subjected to a stream of air, a subcritical bifurcation in convective mean heat transfer occurs at a certain value of angle of attack as control parameter, as shown by Helcig and aus der Wiesche [24]. It was also found that when rotating disks are subjected to a uniform stream of air, a second subcritical bifurcation occurs at transition values depending on both Reynolds numbers and incidence angle of the disk [19, 24].

References

1. John JE, Keith TG (2006) Gas dynamics, 3rd edn. Pearson, Upper Saddle River
2. Lienhard JH, Lienhard JH (2010) A heat transfer textbook. Dover, New York
3. Shevchuk IV (2009) Convective heat and mass transfer in rotating disk systems. Springer, Berlin
4. Schlichting H (1968) Boundary-layer theory. McGraw-Hill, New York
5. Moore FK (1958) On the separation of the unsteady laminar boundary layer. In: Görtler H (ed) Grenzschichtforschung. Springer, Berlin, pp 296–311
6. Rott N (1956) Unsteady viscous flow in the vicinity of a stagnation point. Q J Appl Math 13:444–451
7. Sears WR (1956) Some recent developments in airfoil theory. J Royal Aero Soc 23:490–499
8. Owen JM, Rogers RH (1989) Flow and heat transfer in rotating disc systems, vol 1, Rotor-stator systems. Research Studies, Taunton
9. Owen JM, Rogers RH (1989) Flow and heat transfer in rotating disc systems, vol 2, Rotating cavities. Research Studies, Taunton
10. von Karman T (1921) Über laminare und turbulente Reibung. ZAMM 1:233–252
11. Zandbergen PJ, Dijkstra D (1987) Von Karman swirling flows. Annu Rev Fluid Mech 19:465–491
12. Taylor ES (1974) Dimensional analysis for engineers. Clarendon, Oxford
13. Tobak M, Peake DJ (1982) Topology of three-dimensional separated flows. Annu Rev Fluid Mech 14:61–85
14. Perry AE, Chong MS (1987) A description of eddying motions and flow patterns using critical-point concepts. Annu Rev Fluid Mech 19:125–155
15. Delery JM (2001) Robert Legendre and Henri Werle: towards the elucidation of three-dimensional separation. Annu Rev Fluid Mech 33:129–154
16. Legendre R (1956) Separation de l'ecoulement laminaire tridimensionnel. Rech Aero 54:3–8 (in French)
17. Ligthill MJ (1963) Attachment and separation in three-dimensional flow. In: Rosenhead L (ed) Laminar boundary layers (section II, 2.6). Oxford University Press, New York
18. Andronov AA, Leontovich EA, Gordon II, Maier AG (1973) Qualitative theory of second-order dynamic systems. Wiley, New York
19. Helcig C, aus der Wiesche S, Shevchuk IV (2014) Internal symmetries, fundamental invariants, and convective heat transfer from a rotating disk. In: Proceedings 15th international heat transfer conference (Begell House Digital Library), Kyoto (paper IHTC15-22558)
20. aus der Wiesche S (2004) LES study of heat transfer augmentation and wake instabilities of a rotating disk in a planar stream of air. Heat Mass Transf 40:271–284
21. aus der Wiesche S (2002) Heat transfer and thermal behaviour of a rotating disk passed by a planar air stream. Forschung Ing 67:161–174
22. aus der Wiesche S (2007) Heat transfer from a rotating disk in a parallel air crossflow. Int J Therm Sci 46:745–754
23. Latour B, Bouvier P, Harmand S (2011) Convective heat transfer on a rotating disk with transverse air crossflow. ASME J Heat Transfer. 133, paper-ID 021702 (10 p)
24. Helcig C, aus der Wiesche S (2013) The effect of the incidence angle on the flow over a rotating disk subjected to forced air streams. In: Proceedings ASME fluids engineering summer meeting, Incline Village, Nevada (paper FEDSM2013-16360)

Chapter 3
Wind Tunnel Experiments with Rotating Disks

Investigating convective heat transfer from a rotating disk subjected to streams of air requires placing a heated disk apparatus in the test section of a wind tunnel. This approach was employed for the first time by Dennis et al. [1] in 1970. Although this general experimental approach is fairly straightforward, the performance of accurate measurements is not free of challenges or technical issues. In this section, the approach and procedure are briefly presented and illustrated by examples obtained by the authors and co-workers. Special attention is also given to the importance of the inflow turbulence level because several flow transition and bifurcation phenomena can only be investigated in wind tunnel streams with very low inflow turbulence at the test section.

3.1 Electrically-Heated Disk Approach

Heat transfer coefficients or Nusselt numbers are commonly obtained using two different experimental approaches: (1) the use of an electrically-heated disk and (2) the naphthalene sublimation technique. Whereas the first approach directly yields heat transfer quantities, the second approach permits only the determination of mass transfer quantities, and a further data reduction or recalculation is necessary to obtain corresponding Nusselt numbers. This mass transfer approach was employed for a rotating disk in a parallel stream of air by He et al. [2] in 2005.

Despite the extra calculations and limitations, the naphthalene sublimation technique itself has been used for several decades in the case of rotating disk systems. It yields the Sherwood number Sh as a function of the involved Reynolds number Re and the Schmidt number Sc. A statement about the corresponding heat transfer, i.e., the Nusselt number Nu, can be obtained on the basis of the analogy between heat and mass transfer [3]. This assumption leads to the relation

© The Author(s) 2016
S. aus der Wiesche, C. Helcig, *Convective Heat Transfer From Rotating Disks Subjected To Streams Of Air*, SpringerBriefs in Applied Sciences and Technology, DOI 10.1007/978-3-319-20167-2_3

$$Nu = Sh \, (Pr/Sc)^m. \tag{3.1}$$

Typically, the exponent m in (3.1) is assumed to be constant and is not a function of Pr and Sc. For air at atmospheric conditions, the values $Pr = 0.71$–0.72 and $Sc = 2.28$ for naphthalene are appropriate. In line with the methodology by Cho and Rhee [4], the exponent should be equal to $m = 0.4$, but in the literature, a strong scattering of its values reaching from $m = 0.31$ up to 0.58 can be observed [5, 6]. As discussed by Shevchuk [7], it is obvious that an error in the choice of the value of m can lead to poor results for predictions of the Nusselt number. Furthermore, the uniform and reproducible preparation of the surfaces is not free from difficulties in the naphthalene sublimation technique. In spite of these issues, this method should be carefully considered in cases where absolute statements about Nusselt numbers are required.

The use of approach (1), i.e., an electrically-heated disk and the conduction of temperature and heat transfer measurements, circumvents such issues, but it requires substantial efforts in designing and manufacturing accurate test setups. Such a rotating disk apparatus was successfully used by Cobb and Saunders [8] in 1956. The mean heat transfer coefficients were found directly by measuring the heat input and surface temperature of the disk after steady state had been achieved. The following issues are critical for obtaining high-accuracy data with such apparatus:

1. Uniform or well-defined surface temperature distributions,
2. Elimination of additional heat losses not covered in the data reduction and heat balance calculation,
3. Accurate surface temperature measurements and low uncertainty levels for the data recording system.

An example of such an apparatus used for the experimental research study [9] is shown in Fig. 3.1. The apparatus used for that study consisted of an electrically-heated composite disk with radius $R = 200$ mm carried overhung at one end of a shaft mounted in bearings. The rotordynamics of such an overhung rotor system are in general not trivial [10], and the existence of critical running speeds or gyroscopic effects within the desired range of rotational speeds should be carefully considered for the mechanical concept of the rotating disk test-rig. Lead wires from the heater layer and thermocouples were brought out through slip rings mounted on the shaft. The shaft was driven by an electric motor and was capable of rotational speeds up to 3000 rpm. The test-rig was placed with the disk central in the test section of an open jet wind tunnel with airspeed between $u_\infty = 1.3$ m/s and 40 m/s, enabling translational Reynolds numbers $1.7 \times 10^4 < Re_u = u_\infty \, R/\nu < 5.2 \times 10^5$. To avoid vibrations, the structure was mounted with rubber damping elements to the fixed ground plate.

The disk was a composite structure, as shown in Fig. 3.1 in detail. An electric heater layer foil was glued directly under an aluminum disk with a thickness of 2 mm and a radius of 200 mm. To achieve nearly uniform surface temperatures over the entire disk, the layout of the electric heater layer is important. A powerful tool

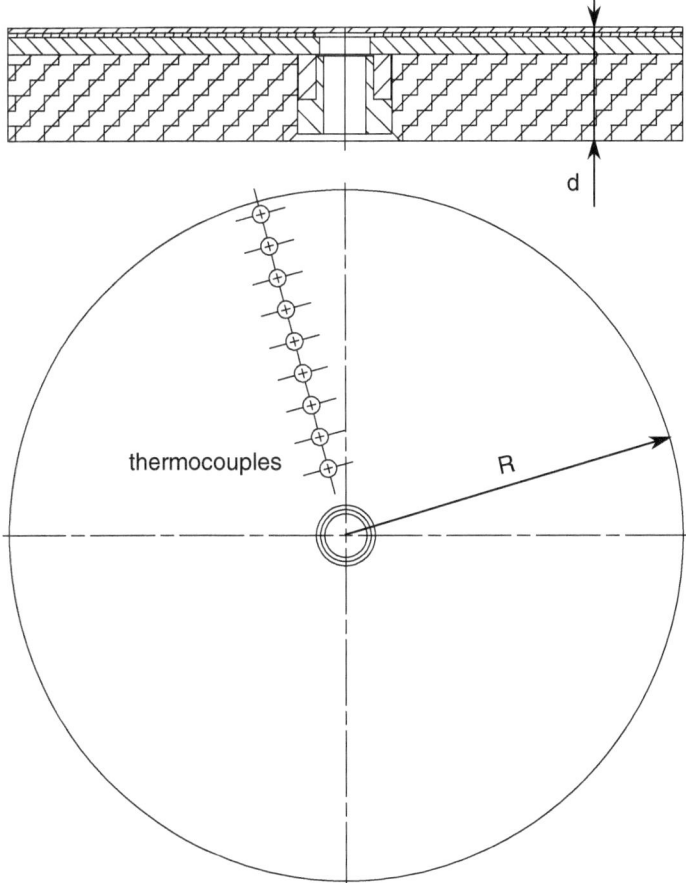

Fig. 3.1 Typical apparatus with electrically-heated disk as used in study [9]

for optimizing it during the design stage can be found in numerical thermal simulations using finite-element methods available in standard commercial software. After manufacturing, the resulting surface temperature distribution should be checked because heat transfer coefficients and mean Nusselt numbers depend on temperature gradients to a certain extent (see Chap. 4 for more details). As an illustrative example, Fig 3.2 shows the measured surface temperature distribution for the disk used in study [9]. For a disk in still air, an isothermal disk surface was almost achieved, as indicated by the IR radiometer plot with a driving temperature difference of order 20 K, see Fig. 3.2.

The upper surface temperature T_w was continuously measured by means of three PT1000 thermocouples (in accordance to DIN EN 60751, class 1/3 DIN) fitted in grooves flush with the surface at different radial locations in [9]. In a later study [11], the number N of thermocouples was increased to a value of $N = 9$. An average surface temperature $T_{w,av}$ can be obtained by

Fig. 3.2 Surface
temperature distribution
for the electrically-heated
disk of Fig. 3.1 as used
in study [9]

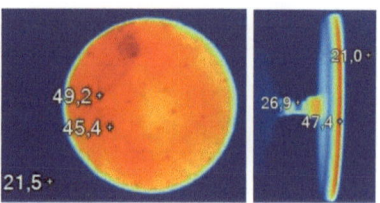

$$T_{w,av} = \sum_{i=1}^{N} g_i T_{w,i} \tag{3.2}$$

with weighting factors

$$g_i = \frac{A_i}{\pi R^2} = \frac{2 r_i \Delta r_i}{R^2} \tag{3.3}$$

for the sensor i placed at a radial coordinate r_i and covering an area A_i. Such a procedure is recommended for calculating the driving temperature difference $\Delta T = T_{w,av} - T_{\infty}$ for the mean heat transfer coefficient h_m. In addition to temperature measurements based on thermocouples, the IR radiometry also offers great potential for heat transfer measurements on rotating disk as demonstrated by Cardone et al. [12]. The exact calibration of the thermoscanner system is however crucial for that method.

Additional heat losses arise due to insufficient thermal insulation of the rim and the backside of the disk, from heat flux from the disk to the shaft, and through radiation and natural convection. The loss due to natural convection will be discussed in detail in Sect. 3.2. Radiation losses are typically of minor importance due to the comparably low temperature level, but natural convection can contribute significantly, particularly in cases with low forced flows. In the early work by Dennis et al. [1], the heat loss from the ring was minimized by fitting a guard rim containing another electric heater around the outside of the main disk and separating it by an insulating ring. The heat flow across the insulating ring was estimated from the temperature difference between two thermocouples, one on each side. This active design can lead to a considerable improvement over the passive method used by Cobb and Saunders [8], who reduced the parasitic heat flow by machining grooves into the contacting faces of the main disk and guard rim minimizing the area of heat conduction. Such a passive mode was also used by Trinkl et al. [9] and in [11]. The heat loss to the shaft can be minimized by including thermal insulation layers between the main disk and the hollow shaft. The resulting parasitic heat loss can be estimated by measuring the temperature difference between two thermocouples mounted on the shaft with an axial distance. In many cases, only one side of the disk is considered as a heat transfer surface, and the backside should be thermally insulated. For composite disk designs, massive insulation layers lead typically to comparably thick disks. In study [9], the final thickness of the composite disk was 24 mm. The parasitic heat losses can reach values of 15 % of the heat input [1]; good designs can reach values below 5 % [9].

A technical difficulty for the measurement system is given by having to send the electric signals of the thermocouples within the rotating disk apparatus to the amplifier and data recording system. In early works [1, 9], slip rings were used. The reproducibility and the accuracy of such an approach are in principle limited due to aging effects of the slip rings, and should be continuously controlled during measurements. In more recent works, the use of telemetry systems has become popular (see, for instance, [11]) because such systems are currently available with small dimensions and at moderate costs.

When the system reaches steady state, the net heat input and the temperatures are measured and the heat balance gives the mean heat transfer coefficient for a heated disk. The time required to achieve a steady state is typically several hours for the disk apparatus described in [1] or [9]. In some cases, e. g., for determining the natural convection contribution, the time required for achieving a statistically steady state is about 30 h. This time period means that a systematic experimental study including several data points occupies a wind tunnel for several weeks or even months. In such cases, automizing the test-setup and the data logger system is recommended.

3.2 Natural Convection Effects

The present book focuses on forced convection effects and convective heat transfer due to forced flows, but for heat transfer experiments with actual disks, natural convection contributions to the observed heat transfer have to be taken into account. Without clarifying the effect of natural convection, experimental data might be misinterpreted. Natural convection plays a dominant role in the case of the very low forced flows that result from very low disk-running speeds or stream velocities, but it is also of some importance even at moderate levels of rotational or translational Reynolds numbers. An excellent discussion and investigation of natural convection effects for rotating disks with and without an outer stream of air has been proposed by Mabuchi et al. [13].

For describing natural convection, the Grashof number

$$Gr = \frac{g\beta_T (2R)^3 \Delta T}{v^2} \tag{3.4}$$

defined by the disk radius R and gravity g can be employed as a suitable dimensionless group. The temperature difference is given for an isothermal surface by means of $\Delta T = T_w - T_\infty$; the bulk modulus of expansion β_T and the physical properties of the fluid can be evaluated at the so-called film temperature $T_f = (T_w + T_\infty)/2$, as discussed in Sect. 3.3 in more detail. Since natural convection has a significant effect in low forced flow velocities, it is possible to limit the discussion to the laminar forced flow regime. The effect of natural convection can

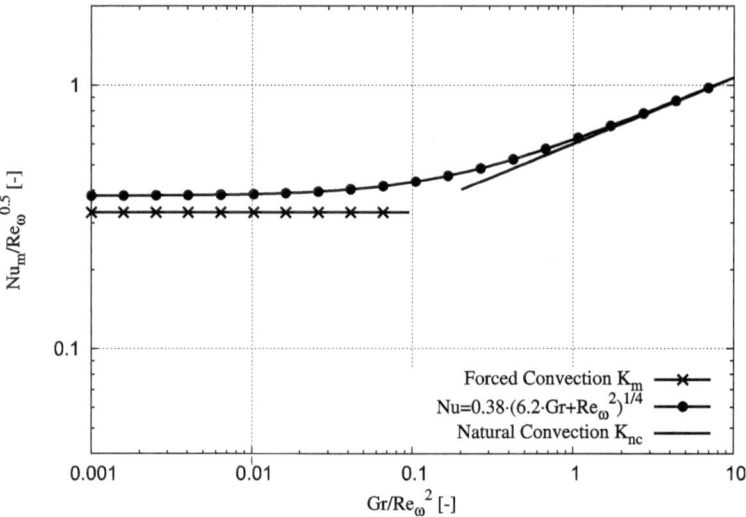

Fig. 3.3 Effect of natural convection on forced convection

be experimentally assessed by means of some measurements of the actual mean Nusselt number $Nu_{m,a}$ for some low forced flow velocities or corresponding Reynolds numbers Re. For a given disk apparatus with radius R, it is useful to plot the obtained $Nu_{m,a}/Re^{1/2}$ against Gr/Re^2 in accordance with the scheme shown in Fig 3.3. For a rotating disk in still air, the Reynolds number is given by means of $Re = Re_\omega = \omega R^2/\nu$ and for a resting disk subjected to a stream of air, the Reynolds number is $Re = Re_u = u_\infty R/\nu$ in Fig. 3.3. It should be noted that sometimes the translational Reynolds number is defined by means of the potential flow expression a R as a characteristic velocity [13]. Then, the corresponding Reynolds number is $Re = Re_a = a R^2/\nu$. For small Grashof numbers in comparison to Re^2, the effect of natural convection is of minor importance, and the experimental data should be in a reasonable agreement with the theoretical curve of forced laminar flow (see line K_m in Fig. 3.3). For large Grashof numbers Gr or small Reynolds numbers Re, the heat transfer is dominated by natural convection, and the experimental data for the actual Nusselt number $Nu_{m,a}$ should be in accordance with a natural convection correlation containing only the Grashof number (see line K_{nc} in Fig. 3.3). For a small *perpendicular* disk at constant surface temperature, Mabuchi et al. [13] recommended the Izumi correlation

$$Nu_{m,nc} = 0.60\,Gr^{1/4} \qquad (3.5)$$

for calculating the mean Nusselt number of natural convection. For other configurations, the reliability of that correlation should be carefully checked using measurements.

Between the limit values $(Gr/Re^2)_1$ and $(Gr/Re^2)_2$, there is a mixed convection regime where both contributions have to be considered. In that regime, an expression

Table 3.1 Values for heat transfer regime limits as recommended by Mabuchi et al. [13] for a perpendicular disk

Limit values	Rotating disk in still air	Resting disk in an air stream
$(Gr/Re^2)_1$	0.01	0.04
$(Gr/Re^2)_2$	2	4

$$Nu_{m,a} = K_m \left(C_{nc} Gr + Re^2 \right)^{1/4} \tag{3.6}$$

with constants K_m and C_{nc} can be employed as an empirical correlation in accordance with Mabuchi et al. [13] for a perpendicular disk. Table 3.1 lists the values for the heat transfer regime limits as recommended by Mabuchi et al. [13]. The limit values in Table 3.1 are recalculated for Re_u for a resting disk in an air stream because Mabuchi et al. used a slightly different definition of the Reynolds number. For a resting disk subjected to a perpendicular stream of air, natural convection effects can be neglected for $Gr \le 0.04\, Re_u^2$. Particularly in the case of small forced flow Reynolds numbers, the effect of natural convection makes an interpretation of experimental data difficult. To evaluate the desired forced flow heat transfer coefficient h_m, the natural convection contribution has to be considered in the energy balance and data reduction.

3.3   Data Reduction for Variable Fluid Properties

In the case of constant fluid properties, data reduction is a straightforward matter and can be performed without any difficulties. For sufficiently large driving temperature differences or high disk surface temperature T_w with regard to the ambient inflow temperature T_∞, the temperature-dependency of the fluid properties has to be taken into account. In the literature, two different methods are common: (1) the property-ratio-method and (2) the reference-temperature method.

In the property-ratio-method, the actual Nusselt number Nu is related to the ideal value Nu_{cp} obtained for constant fluid properties (denoted by index cp) by means of an empirical relation

$$\frac{Nu}{Nu_{cp}} = \left(\frac{\rho_w \mu_w}{\rho_\infty \mu_\infty} \right)^{m_1} \left(\frac{Pr_w}{Pr_\infty} \right)^{m_2} \left(\frac{c_{p,w}}{c_{p,\infty}} \right)^{m_3} \tag{3.7}$$

with empirical exponents m_i. In air with $Pr = 0.71$, relation (3.7) can by replaced by the simple expression

$$\frac{Nu}{Nu_{cp,\infty}} = \left(\frac{T_w}{T_\infty} \right)^{0.004} \tag{3.8}$$

in accordance with Kays and Crawford [14]. Other authors, e. g., [15], recommended a value of 0.02 instead of 0.004 for the exponent in (3.8) for heat transfer with air at atmospheric conditions. Since both values are comparably small, the temperature correction (3.8) is only of minor importance in experiments with a moderately heated disk in air.

In the reference-temperature method, all variables are calculated at an artificial reference temperature T_{ref}, and a further recalculation procedure is applied for data reduction. The choice of the reference temperature T_{ref} is critical. In accordance with Eckert, the so-called film temperature

$$T_{ref} = \frac{T_\infty + T_w}{2} \qquad (3.9)$$

is frequently used.

Both methods are empirically and not rigorously justified in terms of a rational theory. In cases where the above methods lead to considerable differences, the method by Gersten and Herwig [15, 16] based on a perturbation approach is therefore recommended. There, the functions describing the temperature dependence of the fluid properties are expanded as a Taylor series at an arbitrary reference state whose coefficients are dimensionless fluid properties. The results such as the Nusselt number or wall shear stress are formulated as a universal power series of the (small) temperature parameter $(T_w - T_\infty)/T_\infty$. This rational method has the great advantage that the empirical character of the other methods is circumvented. Furthermore, Gersten and Herwig's analytical treatment permits the values for the exponents in (3.7) to be determined, too.

3.4   Wind Tunnel and Inflow Turbulence

In many experiments for convective heat transfer effects, the quality of the stream of air generated by the wind tunnel is not crucial, and comparably simple wind tunnels can be used. For rotating disks with finite thickness, some flow and heat transfer regime transitions can only be observed in stream of airs with fairly low inflow turbulence over the entire test section. This should be catered for by the selection of the wind tunnel.

A wind tunnel is a tool used in research to study the effects of air moving past solid objects. It consists of a test section with the object under test mounted in the middle. Air is moved past the object using a powerful fan system. Reviews of the many types of wind tunnels, which include low-speed, high-speed, intermittent blow-down, and suction tunnels, are available in the literature, e. g., Barlow et al. [17]. For the present purposes, only low-speed wind tunnels operating in the range of 1–60 m/s for the inflow speed are required.

A typical low-speed, continuous running wind tunnel with an open test section is shown in Fig. 3.4. Such a facility was used in the studies [9] and [11]. The velocity profiles at the test section where the disk apparatus was placed are plotted in Fig. 3.5 for that wind tunnel. The core radius of the circular test section was 240 mm. The use of an open jet wind tunnel has the advantage that blockage effects due to

Fig. 3.4 Low-speed wind tunnel as used in studies [9] and [11]

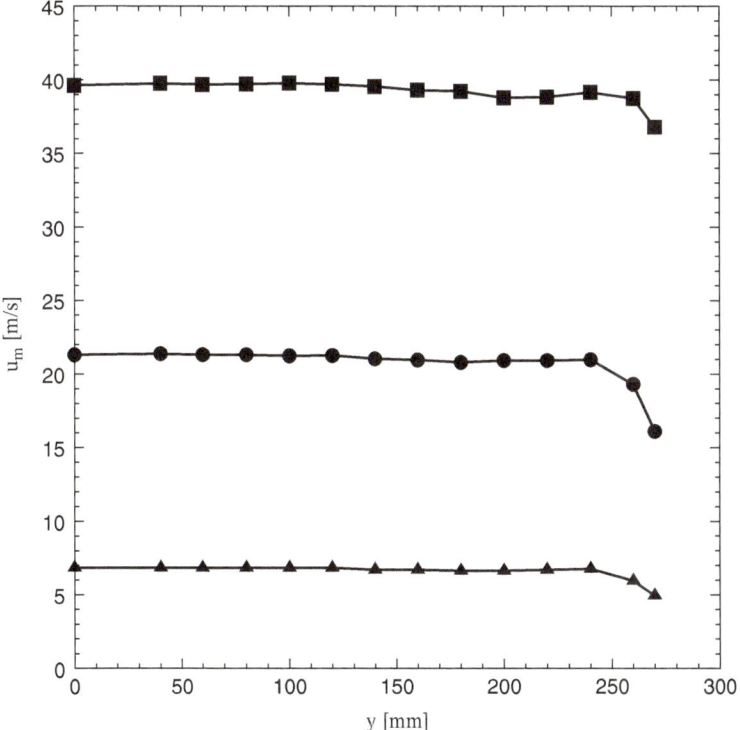

Fig. 3.5 Measured crossflow velocity profile of the wind tunnel of study [9] at the test section

the test object are not as extreme as in closed test sections. In cases where blockage effects are likely to occur, correction functions for the wind tunnel results are recommended in order to estimate the "infinite" inflow configuration. The results shown in Fig. 3.5 demonstrate that a nearly uniform inflow velocity u_∞ was reached within the test section for a broad range of speed levels.

Variations in flow quality between two different wind tunnels or experimental studies will cause variations between the results obtained from the two test facilities when identical experiments are performed. Such variations are extremely significant in cases where transition phenomena from laminar to turbulent flow are being studied. In addition to the mean or averaged wind tunnel speed u_∞, the inflow turbulence has to be specified as well. Since Reynolds, the velocity u of a turbulent flow can be separated into a time-average value u_∞ and a fluctuating turbulent contribution u'. The turbulence level Tu of a wind tunnel is given in the case of isotropic turbulence by the simple expression

$$\text{Tu} = \frac{\sqrt{u'^2}}{u_\infty}. \qquad (3.10)$$

It can be measured directly by hot-wire anemometry or indirectly by determining the so-called turbulence factor TF of a sphere.

Spheres are known to have a distinct critical Reynolds number above which the flow on the upstream face of the sphere is fully turbulent causing the drag coefficient to drop dramatically [17]. The Reynolds number at which this transition occurs is strongly dependent on the degree of turbulence in the wind tunnel. It is known that the drop in drag due to that transition for a perfectly smooth sphere in an atmospheric flow will occur at $Re_{d,cr,\infty} = 385,000$ [17]. The concept of the turbulence factor TF relates the actual transition Reynolds number $Re_{d,cr} = u_\infty d/\nu$ of the sphere with diameter d to that ideal value, i.e.,

$$\text{TF} = \frac{385,000}{Re_{d,cr}}. \qquad (3.11)$$

The actual transition Reynolds number $Re_{d,cr}$ is the Reynolds number at which the measured drag coefficient C_D passes through 0.3 during the transition from laminar to turbulent boundary layer flow. This is illustrated in Fig. 3.6 by means of data obtained for the wind tunnel used in study [9]. The object of the tests was a smooth bowling sphere. The drag force F_D is related to the drag coefficient C_D in accordance with

$$F_D = C_D \frac{1}{2} \rho u_\infty^2 \frac{\pi}{4} d^2. \qquad (3.12)$$

The experiments fitted well to a polynomial curve, as shown in Fig. 3.6, yielding a turbulence factor of TF $= 1.16 \pm 0.01$, which was a reasonable value for such an open jet wind tunnel facility. During the testing, it was typically observed that the amount of vibration in the mounting structure decreased as the transitional region was surpassed. For poor mounting structures, the mechanical vibrations due to aerodynamic excitation forces might lead to high noise-to-signal ratios making the transition very difficult to determine.

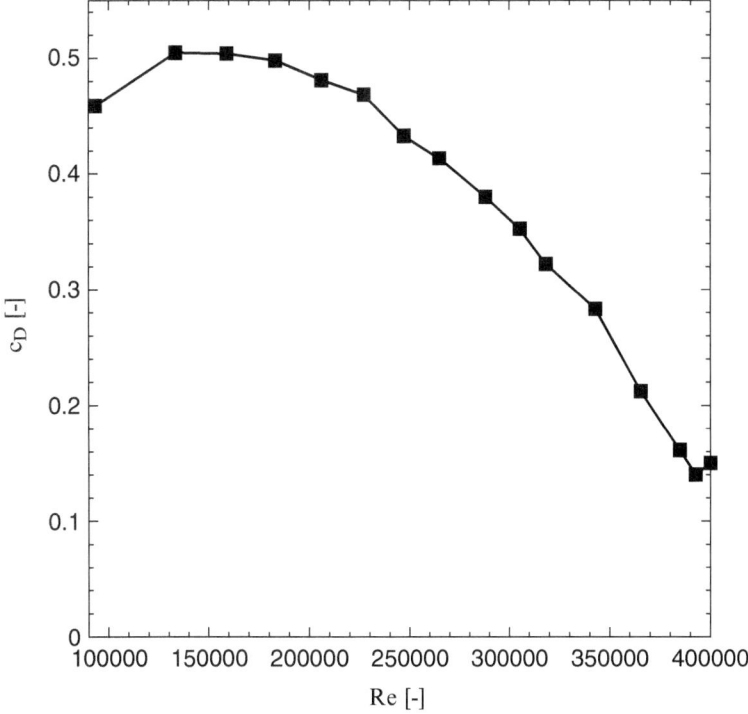

Fig. 3.6 Measured drop in drag for a sphere placed in the test section of the low-speed wind tunnel as used in study [9]

It should be noted that some authors [18] define the transition through a vanishing pressure difference between the upstream side of the sphere and the static inflow pressure. The corresponding ideal Reynolds number for that transition is of order 4.1×10^5, and then the definition of such a turbulence factor differs obviously from (3.5).

The turbulence factor TF can be related to the wind tunnel turbulence level using the classic relation obtained in the hot-wire anemometry by Dryden and Keuthe [19]. For low turbulence levels, the simple linear relation

$$\frac{\text{Tu}}{[\%]} = 1.25 \times (\text{TF} - 1.00) \tag{3.13}$$

might be appropriate for relating the turbulence level Tu (in % turbulence) to the turbulence factor TF defined by (3.11). A turbulence factor of TF $= 1.16$ corresponds to a turbulence level of Tu $= 0.2$ %.

References

1. Dennis RW, Newstead C, Ede AJ (1970) The heat transfer from a rotating disc in an air crossflow. In: Proceedings of 4th international heat transfer conference, Versailles, 1970 (paper FC 7.1)
2. He Y, Ma L, Huang S (2005) Convection heat and mass transfer from a disk. Heat Mass Transf 41:766–772
3. Baehr HD, Stephan K (1996) Wärme- und Stoffübertragung. Springer, Berlin
4. Cho HH, Rhee DH (2001) Local heat/mass transfer measurements on the effusion plate in impingement/effusion cooling systems. ASME J Turbomachinery 123:601–608
5. Cho HH, Won CH, Ryu GY, Rhee DH (2002) Local heat transfer characteristics in a single rotating disk and co-rotating disks. Microsyst Technol 9:399–408
6. Shimada R, Naito S, Kumagai S, Takeyama T (1987) Enhancement of heat transfer from a rotating disk using turbulence promoter. JSME Int J Ser B 30:1423–1429
7. Shevchuk IV (2009) Convective heat and mass transfer in rotating disk systems. Springer, Berlin
8. Cobb EC, Saunders OA (1956) Heat transfer from a rotating disc. Proc R Soc A 236:343–351
9. Trinkl CM, Bardas U, Weyck A, aus der Wiesche S (2011) Experimental study of the convective heat transfer from a rotating disc subjected to forced air streams. Int J Thermal Sci 50:73–80
10. Childs D (1993) Turbomachinery rotordynamics. Wiley, New York
11. Helcig C, aus der Wiesche S (2013) The effect of the incidence angle on the flow over a rotating disk subjected to forced air streams. In: Proceedings ASME fluids engineering summer meeting, Incline Village, Nevada (paper FEDSM2013-16360)
12. Cardone G, Astarita T, Carlomagno GM (1997) Heat transfer measurements on a rotating disk. Int J Rotating Machinery 3:1–9
13. Mabuchi I, Tanaka T, Sakakibara Y (1971) Studies on the convective heat transfer from a rotating disk (5th report, experiment on the laminar heat transfer from a rotating isothermal disk in a uniform forced stream). Bull JSME 14:581–589
14. Kays WM, Crawford ME (1980) Convective heat and mass transfer. McGraw-Hill, New York
15. Schlichting H, Gersten K (1997) Grenzschicht-Theorie, 9th edn. Springer, Berlin
16. Gersten K, Herwig H (1984) Momentum and heat transfer for the laminar flat-plate flow with variable fluid properties. Wärme- und Stoffübertragung (Heat Mass Transfer) 18:25–35 (in German)
17. Barlow JB, Rae WH, Pope A (1999) Low speed wind tunnel testing. Wiley, New York
18. See, for instance: AVA Monographie D1 Modellversuchstechnik 1, Göttingen 1946, D1 4.2 (in German)
19. Dryden HL, Keuthe AM (1929) Effect of turbulence in wind tunnel measurements. NACA Report 342

Chapter 4
Axisymmetric Configurations

For axisymmetric configurations of rotating disk systems, fluid flow and heat transfer can be analyzed using powerful mathematical methods such as the self-similar solutions. These solutions cover the free rotating disk without any additional forced flow and the rotating disk placed perpendicular to a uniform stream. They represent important limit cases for the more general configuration of an inclined rotating disk, and since there are a lot of reliable data for these cases, they are useful for validating experimental or numerical methods. However, sometimes the analytical treatment is only valid under certain conditions that might not be fulfilled in practical applications, for example, in the case of a stagnation flow onto an orthogonal disk. A closed self-similar solution is only known for an infinite disk, and its application to finite disks in uniform streams still has to be confirmed.

4.1 Free Rotating Disk in Still Air

Heat transfer from free rotating disk can be calculated using a *local* Nusselt number

$$Nu = \frac{\dot{q}_w r}{\lambda(T_w - T_\infty)} = K \cdot Re_{\omega,r}^{n_R} \quad \text{with} \quad Re_{\omega,r} = \frac{\omega r^2}{\nu} \tag{4.1}$$

defined by means of local variables and a radial coordinate r for the disk with angular velocity ω, or by a *mean* or *average* Nusselt number

$$Nu_m = \frac{h_m R}{\lambda} = K_m \cdot Re_\omega^{n_R} \quad \text{with} \quad Re_\omega = \frac{\omega R^2}{\nu} \tag{4.2}$$

S. aus der Wiesche, C. Helcig, *Convective Heat Transfer From Rotating Disks Subjected To Streams Of Air*, SpringerBriefs in Applied Sciences and Technology, DOI 10.1007/978-3-319-20167-2_4

defined by means of the outer disk radius R as the length scale and the mean heat transfer coefficient h_m over the entire disk [1–4]. The coefficients K and K_m in (4.1) and (4.2) depend on the flow regime, the Prandtl number Pr, and the temperature distribution on the surface, while exponent n_R depends only on the flow regime. Numerous research studies have investigated this topic for a free rotating disk, as reviewed by Shevchuk [4], but experimental data for K for nonisothermal surfaces and other values of Prandtl number over the range $Pr \leq 1$ are not yet available.

For *laminar* flow over the rotating disk, the local and the mean constants are equal, $K = K_m$, and the exponent is $n_R = 1/2$. Theoretical values of the constant K can be found by means of the self-similar solution following the von Karman swirl flow approach (see Chap. 2). In detail, using the notation of Chap. 2 for $B = \Omega/\omega = 0$ and $N = u_{r,\infty}/\omega \ r = 0$ and the following expressions for the velocity field provides

$$F(\zeta) = \frac{u_r}{\omega r}, \quad H(\zeta) = \frac{u_z}{\sqrt{\omega r}}, \quad G(\zeta) = \frac{u_\phi}{\omega r} \quad \text{with} \quad \zeta = z\sqrt{\frac{\omega}{v}}. \tag{4.3}$$

With the normalized temperature

$$\Theta = \frac{T - T_\infty}{T_w - T_\infty}, \tag{4.4}$$

the energy equation is

$$\Theta'' - Pr\left(n^*F\Theta + H\Theta'\right) = 0, \tag{4.5}$$

for a surface temperature distribution

$$T_w = T_0 + C r^{n^*}. \tag{4.6}$$

The boundary conditions for the ordinary differential equation (4.5) are

$$\zeta = 0: \quad \Theta = 1 \quad \text{and} \quad \zeta \to \infty: \quad \Theta = 0. \tag{4.7}$$

Together with the ordinary differential equations for the flow field, see Chap. 2, the above set of equations can be solved. From a mathematical point of view, the set of ordinary differential equations represents an exact solution of the Navier–Stokes equation. The constant K can be found by means of the identity

$$K = -\Theta'\big|_{\zeta=0}. \tag{4.8}$$

Some profiles for velocity and temperature (for $n^* = 0$) are shown in Fig. 4.1.

For laminar flow over a rotating disk, the value of the exponent n^* in the boundary condition (4.6) is the same $n^* = 0$ for an isothermal surface $T_w = $ constant

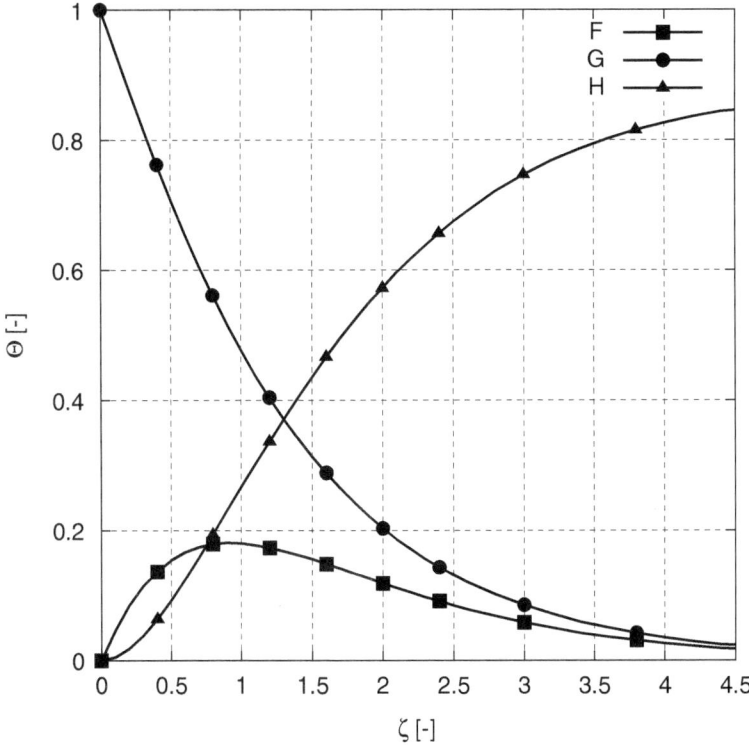

Fig. 4.1 Velocity and temperature profiles in laminar flow over a free rotating disk in still air

Table 4.1 Values of the constant K according to the exact solution for laminar flow over a free rotating disk in a still fluid (data from [4])	Pr	$n^* = -1$	$n^* = 0$	$n^* = 1$
	1.0	0.2352	0.3963	0.5180
	0.8	0.2046	0.3495	0.4608
	0.71	0.1893	0.3259	0.4319
	0.6	0.1691	0.2943	0.3929
	0.4	0.1267	0.2263	0.3078
	0.2	0.0732	0.1362	0.1912

and a constant heat surface heat flux. Some numerically obtained values for the constant K are listed in Table 4.1. The most complete and accurate data are provided by Shevchuk [4]. The analytical treatment predicts for air at atmospheric condition (Pr of order 0.71 up to 0.72) a value $K = K_{\mathrm{m}} = 0.32 - 0.34$ (see Table 4.1). That value is in reasonable agreement with the most reliable experimental data [5, 6] for that configuration. In early works [7, 8], slightly higher values were reported of order $K - 0.38 - 0.4$. A recent study [9] with an apparatus designed for a test section of a wind tunnel yielded $K = 0.36$ for the laminar flow regime over a free rotating disk in still air with a nearly isothermal surface.

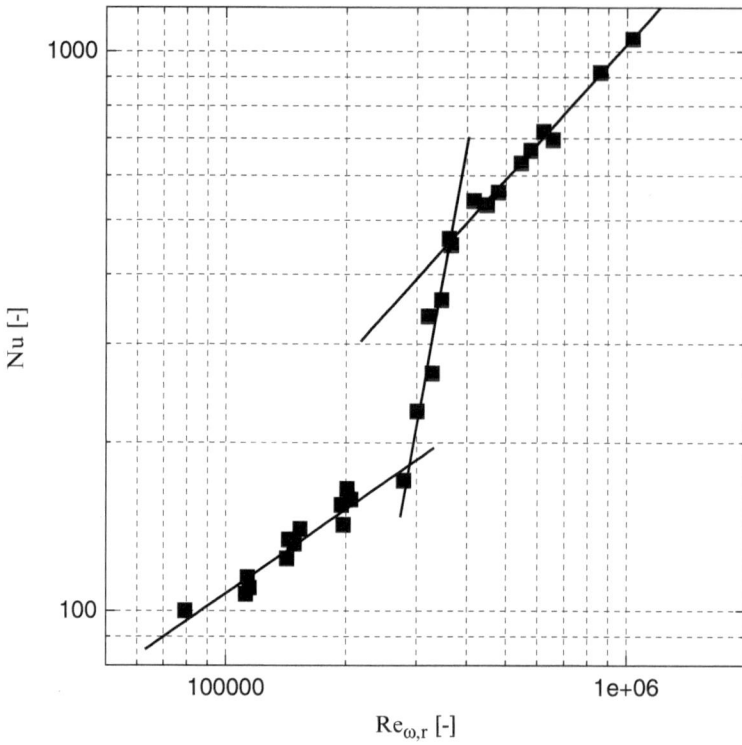

Fig. 4.2 Local Nusselt numbers of a rotating disk in still air (adapted from [4])

The *transition* from laminar to turbulent flow over a rotating disk is accompanied by instability of laminar flow, the occurrence of spiral vortices, and a further development of turbulence. The classic work on this kind of hydrodynamic instability was written by Gregory, Stuart, and Walker [10] in 1956, but a lot of other studies about this topic are also available as discussed by Shevchuk [4]. The transition leads to a fairly sharp increase of the local Nusselt number Nu and to a substantial increase of the mean Nusselt number Nu_m for rotational Reynolds numbers Re_ω larger than a critical value of order $Re_\omega = 240{,}000$. The heat transfer augmentation due to turbulent flow is demonstrated by means of Figs. 4.2 and 4.3 where experimentally determined local and the mean Nusselt numbers are shown as functions of the rotational Reynolds numbers. In literature, different values for the onset of the transition and for the end of the transition are given. Values of order $Re_{\omega,r} = 8.6 \times 10^4 - 2.1 \times 10^5$ are stated for the onset of the transition [11, 12]. The theoretical value based on a linear perturbation approach is $Re_{\omega,r} = 8.17 \times 10^4$, which is in reasonable agreement with experiments [11]. For a sufficiently large rotational Reynolds number, the end of the transition and the fully turbulent flow regime are achieved. Values of order $Re_{\omega,r} = 2.5 \times 10^5 - 3.6 \times 10^5$ are stated for the end of the transition [13, 14].

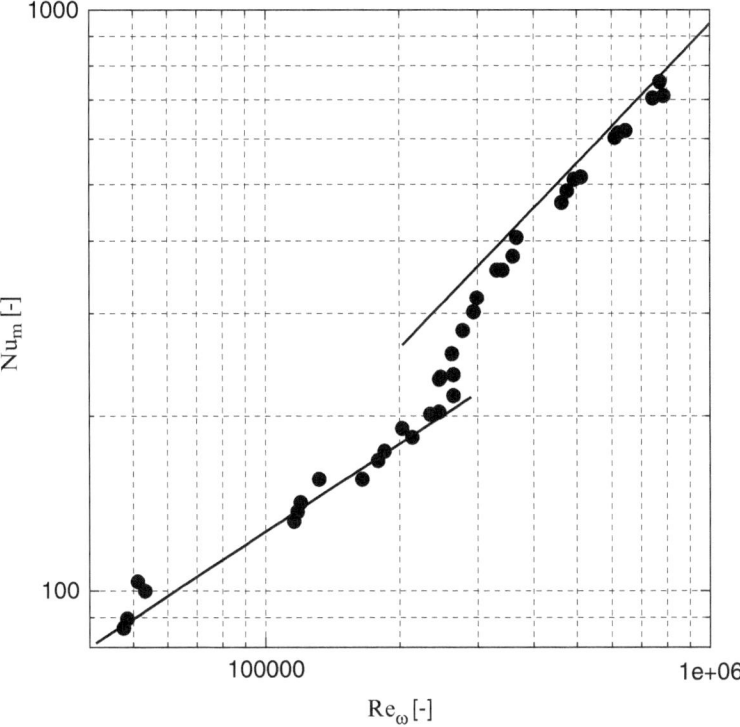

Fig. 4.3 Mean Nusselt numbers of an entire rotating disk in still air (adapted from [4])

The *fully turbulent flow* regime is reached for a rotational Reynolds number of order $Re_\omega = 10^6$, as is illustrated by Fig. 4.3. Then, the mean Nusselt number Nu_m can be well described by a correlation (4.2) with $K_m = 0.015$ and exponent $n_R = 0.8$ [7, 9]. Notably, even in such a case, a large area of the rotating disk is still within the laminar flow regime, but the strong turbulent heat transfer mechanism is dominant for the mean Nusselt number Nu_m. This result can be seen in Fig. 4.3 by means of the agreement of the experimental data points with the asymptotic correlation developed for fully turbulent flow. The exponent $n_R = 0.8$ is characteristic of the turbulent flow regime. The local and the mean constants, K and K_m, are, in general, not identical for turbulent flow. Using an integral method, Dorfman [1] obtained the following relations

$$K = 0.0197 \, (n^* + 2.6)^{0.2} Pr^{0.6} \quad \text{and} \quad K_m = K \frac{n^* + 2}{n^* + 2.6}. \qquad (4.9)$$

The experimental data for an isothermal surface are in reasonable agreement with that prediction, but data obtained from more recent investigations [4] are slightly lower than the values predicted by (4.9).

The average Nusselt number Nu_m for an entire disk, see Fig. 4.3, can be calculated by dividing the disk surface into a laminar, a transitional, and a fully turbulent zone. In many studies, a fairly simple model of only two zones is applied. Then, the transition value of the Reynolds number remains as a fitting parameter with a typical value of order 240,000.

4.2 Stationary Disk in an Orthogonal Stream of Air

The convective heat and mass transfer from a stationary disk with radius R placed in an orthogonal stream of air has been investigated in greater detail [15, 16]. Since the disk is stationary, only the translational Reynolds number Re_u and the Prandtl number Pr are involved. The empirical correlation

$$Nu_m = \frac{h_m R}{\lambda} = 0.63 \cdot Re_u^{1/2} \qquad \text{for} \quad Pr = 0.71 \tag{4.10}$$

has proven its reliability for a large range of Reynolds numbers $Re_u = u_\infty R/v$ for this configuration of an isothermal disk surface.

In an experimental study by Mabuchi et al. [17], a correlation

$$Nu_m = \frac{h_m R}{\lambda} = 0.70 \cdot Re_a^{1/2} \qquad \text{for} \quad Pr = 0.71 \tag{4.11}$$

with a value of 0.70 was obtained. This value agrees well with the theoretical prediction 0.669 based on a self-similar solution method. Mabuchi et al. [17] defined their Reynolds number $Re_a = aR^2/v$ using the potential flow expression aR as the characteristic velocity. With the identity $Re_u = (\pi/2)\, Re_a$, it follows that their findings would correspond to a constant of 0.559 in (4.10) instead of 0.63. The discrepancy between the values for the mean Nusselt number can be explained as follows. Mabuchi et al. [17] investigated the heat transfer behavior under the assumption of the validity of a self-similar solution. Substantial theoretical efforts are, however, required if the effect of flow separation at the outer rim of the disk with finite radius R is to be accounted for accurately. In the case of an infinite disk, an analytical self-similar solution of the Navier–Stokes equation is known [18], and the local Nusselt number can be calculated based on that solution. But once separation occurs at the outer rim of a finite disk, the velocity component in the radial direction is accelerated at the neighborhood of the outer part of the disk. As a consequence, the local heat transfer coefficient rises there. Therefore the self-similar solution yields noticeably lower values for the constant in (4.10). A detailed discussion of the corresponding boundary layer theory and experimental results can be found in [16]. That study also contains an investigation of the effect of inflow turbulence. In line with the notation of the present book, the correlation

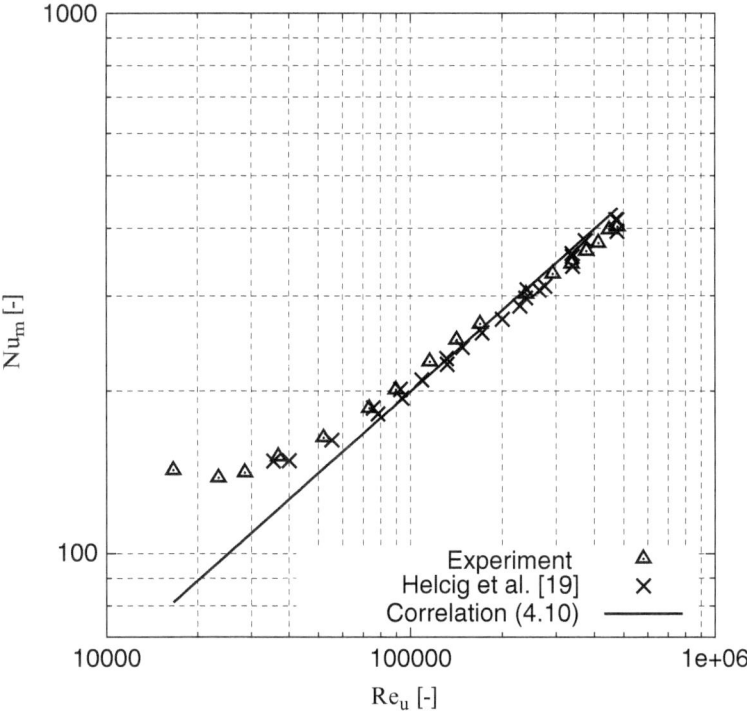

Fig. 4.4 Mean Nusselt numbers of an entire stationary disk subjected to an orthogonal stream of air

$$\frac{Nu(r=0)}{\sqrt{Re_u}} = 0.484 + 0.0247 \frac{Tu}{100} \sqrt{Re_u} \qquad \text{for} \quad Pr = 0.71 \qquad (4.12)$$

for the local Nusselt number $Nu = h(r=0) \, R/\lambda$ at the stagnation point at the disk center $(r=0)$ was derived in study [16] by means of the measurements with different turbulence levels Tu (in %) in the range $Tu\sqrt{Re_u}/100 < 10.6$. Due to the accelerated outer flow at the disk rim, the local Nusselt number $Nu(r)$ increases with increasing radial coordinate r. An integration of the local Nusselt number $Nu(r)$ over the radial coordinate r yields the mean Nusselt number Nu_m.

For a two-dimensional stagnation point (e.g., at a long cylinder), the local Nusselt number decreases with increasing distance from the stagnation point. Then, the maximum value is in given by a correlation $Nu(0) = 0.701 \, (Re_u)^{1/2}$ at the two-dimensional stagnation point in accordance with Eckert [16]. The literature also shows that both the radius R and the diameter are used for defining Nusselt and Reynolds numbers.

For sufficiently low Reynolds numbers, the laminar axisymmetric stagnation flow onto the circular disk is very robust and hence these numbers are frequently used for calibration. Whereas the early studies [15, 16] considered comparably low Reynolds number levels, the experimental data shown in Fig. 4.4 demonstrate that the laminar flow regime can be extended to large Reynolds number levels if low turbulence levels of the uniform stream are given [19]. The data has been obtained by using the disk apparatus described in more detail in Chap. 3. Due to the high Reynolds number level, natural convection was neglected in data reduction.

4.3 Rotating Disk and an Orthogonal Uniform Flow

For an *infinite* rotating disk, the self-similar solution method is still applicable, see Chap. 2. Then, the external flow is given by a potential flow far away from the infinite disk in accordance with

$$\zeta \to \infty: \quad u_{r,\infty} = ar, \quad u_{z,\infty} = -2az, \quad u_{\phi,\infty} = 0 \qquad (4.13)$$

with a potential flow constant a. In any practical application, the disk has a *finite* disk radius R, and it is common to assume the relation

$$a = \frac{2}{\pi R} u_{\text{jet}} = \frac{2u_\infty}{\pi R} \qquad (4.14)$$

with a jet velocity u_{jet} as illustrated in Fig. 4.5. The stagnation pressure p_s on a disk at local radius r can then be written as

$$p_s = p(r) + \frac{1}{2}\rho u_{r,\infty}^2 \qquad (4.15)$$

with (4.13) and (4.14) for the far-field velocity in (4.15).

The common practice of applying the self-similar approach to a disk with a finite radius can lead to serious errors if the theoretical limits of that approach are not appreciated. Firstly, the pressure distribution (4.15) agrees well for radial distances up to $r/R = 0.85$ [17], but it disagrees for larger radial distances due to effects of

Fig. 4.5 Flow and heat transfer between a rotating disk and a perpendicular uniform flow perpendicular (adapted from [4])

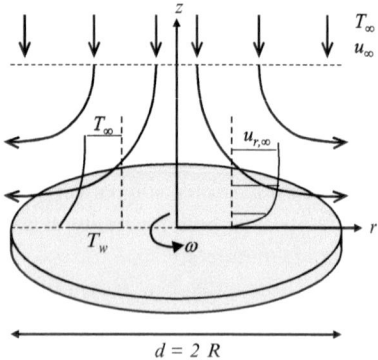

flow separation at the outer rim of the finite disk. As a result, the self-similar approach based on the model assumption of an outer potential flow accurately predicts convective heat transfer only for sufficiently small regions around the stagnation point. A detailed study [17] conducted by Mabuchi and co-workers showed that the self-similar solution for the Nusselt number Nu agrees fairly well with experimental data within the inner area of the disk $r/R < 0.8$.

The calculation of the mean Nusselt number for a disk with finite radius is rather complicated, because the constant a becomes a function of the radial coordinate r, and the constant K_{2R} of the mean Nusselt number correlation

$$Nu_{2R} = K_{2R} \cdot (Re_{\omega,2R} + Re_{a,2R})^{1/2} \qquad (4.16)$$

depending on both Reynolds numbers

$$Re_{\omega,2R} = \frac{\omega (2R)^2}{\nu} \quad \text{and} \quad Re_{a,2R} = \frac{a (2R)^2}{\nu} \qquad (4.17)$$

increases with increasing constant a. In the above definitions, the diameter $2R$ has been used as the length scale, in accordance with the notation of [4]. A correlation could be also formulated by means of a single combined Reynolds number

$$Re_e = \sqrt{a^2 + \omega^2} \frac{R^2}{\nu}, \qquad (4.18)$$

and the mean Nusselt number can be correlated in the laminar flow regime by means of

$$\frac{Nu_m}{\sqrt{Re_e}} = f(a/\omega) \qquad (4.19)$$

with the function f depending on the ratio a/ω. The Prandtl number is fixed to $Pr = 0.71$ in the following. Values for f can be found numerically on the basis of a self-similar solution approach [17]. Some values for f are listed in Table 4.2 for an isothermal disk. The table additionally contains the corresponding Reynolds number ratio Re_ω/Re_u. The limit case $a/\omega = 0$ corresponds to the rotating disk in still air (see Sect. 4.1 and Table 4.1) whereas the case of a resting disk in a stream of air is achieved for $a/\omega \to \infty$ (see Sect. 4.2).

Table 4.2 Values for function f in correlation (4.19) and Reynolds number ratio in accordance with [17]

a/ω	0	0.1	0.25	0.5	1	2	∞
f	0.329	0.360	0.412	0.488	0.580	0.640	0.669
Re_ω/Re_u	∞	6.366	2.546	1.273	0.637	0.318	0

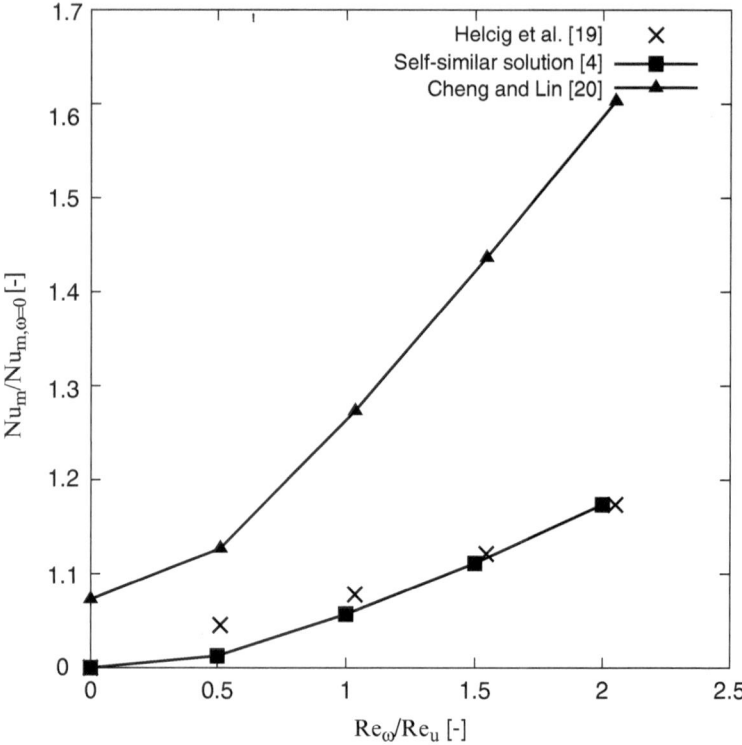

Fig. 4.6 Mean Nusselt number of a rotating disk in a stream of air for a fixed translational Reynolds number (adapted from [17])

Figure 4.6 shows experimental data for a rotating isothermal disk with finite radius R subjected to a uniform stream of air with inflow velocity u_∞. The mean Nusselt number Nu_m of the rotating disk is related to its stationary value $Nu_{m,\omega=0}$ obtained for a stationary perpendicular disk (see Sect. 4.2). Figure 4.5 reveals the increase of the mean Nusselt number with increasing rotational Reynolds number $Re_\omega = \omega R^2/\nu$ related to the fixed translational Reynolds number $Re_u = u_\infty R/\nu$. In addition to the experimental data obtained for the apparatus described in Chap. 3, the results of the analytical treatment based on the self-similar solution have been plotted in Fig. 4.6.

With correlation (4.19) and the values for the function f of Table 4.1, the theoretical increase

$$\frac{Nu_m}{Nu_{m,\omega=0}} = \frac{h_m}{h_{m,\omega=0}} = \frac{f(a/\omega)}{0.669}\left(1 + \frac{\omega^2}{a^2}\right)^{1/4} \tag{4.20}$$

of heat transfer due to rotation can be easily obtained under the assumption of validity of the self-similar solution. The value 0.669 in (4.20) corresponds to the self-similar solution for the resting disk in an orthogonal stream of air (see Table 4.2). The agreement is good between the experimental data and the predictions of the analytical treatment for higher Reynolds number ratios. Due to the flow separation effects at the outer rim of the finite disk, the self-similar solution is slightly worse in the case of dominant stagnation flow, where the measured data for the heat transfer from the entire disk are slightly higher. However, an oversimplified correlation approach [20] with a combined Reynolds number can lead to substantial errors as indicated by the data shown in Fig. 4.6. It is certainly remarkable that, even in the case of high rotational Reynolds numbers, the heat transfer augmentation due to rotation is weak in comparison to the forced stream. In contrast, a significant heat transfer augmentation can be achieved due to an additional stream of air in the opposite case of a rotating disk in still air. Even in cases where the translational Reynolds number Re_u is much smaller than the rotational Reynolds number Re_ω, a noticeable increase of the mean Nusselt number Nu_m results.

4.4 Jet Impingement onto an Orthogonal Rotating Disk

Jet impingement onto an orthogonal surface is of great technical importance for cooling and drying technology. Typically, small jets with diameter d_{jet} are directed onto a solid surface at distance h_{jet}. For stationary flat plates, a very large number of studies and review papers are available, e.g. [21–23]. For rotating disk, the reader is referred to the corresponding chapter of the monograph [4]. In the present contribution, only a uniform stream of air is considered and the interesting phenomena due to confined jets are excluded.

References

1. Dorfman LA (1963) Hydrodynamic resistance and the heat loss of rotating solids. Oliver & Boyd, Edinburgh
2. Owen JM, Rogers RH (1989) Flow and heat transfer in rotating disc systems, vol 1, Rotor-stator systems. Research Studies Press Ltd., Taunton
3. Owen JM, Rogers RH (1989) Flow and heat transfer in rotating disc systems, vol 2, Rotating cavities. Research Studies Press Ltd., Taunton
4. Shevchuk IV (2009) Convective heat and mass transfer in rotating disk systems. Springer, Berlin
5. Eaton JK (1996) Structure and modeling of the three dimensional boundary layer on a rotating disk. Final Report to U.S. Department of Energy, Grant Number DE-FG03-93ER14317-A000, Stanford University
6. Elkins CJ, Eaton JK (2000) Turbulent heat and momentum transport on a rotating disk. J Fluid Mech 402:225–253

7. Dennis RW, Newstead C, Ede AJ (1970) The heat transfer from a rotating disc in an air crossflow. In: Proceedings of 4th International Heat Transfer Conference, Paris-Versailles, 1970 (paper FC 7.1)

8. Booth GL, de Vere APC (1974) Convective heat transfer from a rotating disc in a transverse air stream. In: Proceedings of 5th International Heat Transfer Conference, Tokyo, 1974 (paper FC1.7)

9. Trinkl CM, Bardas U, Weyck A, aus der Wiesche S (2011) Experimental study of the convective heat transfer from a rotating disc subjected to forced air streams. Int J Therm Sci 50:73–80

10. Gregory N, Stuart JT, Walker WS (1956) On the stability of the three-dimensional boundary layers with application to the flow due to a rotating disk. Phil Trans Roy Soc A 248:155–199

11. Malik MR, Wilkinson SP, Orszag SA (1981) Instability and transition in a rotating disc. AIAA J 19:1131–1138

12. Smith NH (1946) Exploratory investigation of laminar boundary layer oscillations on a rotating disc. NACA Tech. Note 1227: 1–21

13. Popiel CO, Boguslawski L (1975) Local heat-transfer coefficients on the rotating disk in still air. Int J Heat Mass Transf 18:167–170

14. Elkins CJ, Eaton JK (1997) Heat transfer in the rotating disk boundary layer. Thermosciences Division Report TSD-103, Stanford University

15. Beg SA (1973) Forced convective mass transfer from circular disks. Wärme- und Stoffübertragung (Heat Mass Transfer) 1:45–51

16. Kottke V, Blenke H, Schmidt KG (1977) Messung und Berechnung des örtlichen und mittleren Stoffüberganges an stumpf angeströmten Kreisscheiben bei unterschiedlicher Turbulenz. Wärme- und Stoffübertragung (Heat Mass Transfer) 10:89–105

17. Mabuchi I, Tanaka T, Sakakibara Y (1971) Studies of convective heat transfer from a rotating disk (5th Report, Experiment on the laminar heat transfer from a rotating isothermal disk in a uniformed forced stream). Bull JSME 14:581–589

18. Schlichting H, Gersten K (1997) Grenzschicht-Theorie, 9th edn. Springer, Berlin

19. Helcig C, aus der Wiesche S (2013) The effect of the incidence angle on the flow over a rotating disk subjected to forced air streams. In: Proceedings ASME Fluids Engineering Summer Meeting, Incline Village, Nevada (paper FEDSM2013-16360)

20. Cheng W, Lin H (1994) Unsteady and steady mass transfer by laminar forced flow against a rotating disk. Heat Mass Transf 30:101–108

21. Martin H (1977) Heat and mass transfer between impinging gas jets and solid surfaces. Adv Heat Transf 13:1–60

22. Downs SJ, James EH (1987) Jet impingement heat transfer—a literature survey. ASME paper 87-HT-35

23. Jambunathan EL, Lai E, Moss MA, Button BL (1992) A review of heat transfer data for single circular jet impingement. Int J Heat Fluid Flow 13:106–115

Chapter 5
Stationary Disk in Air Stream

No general solution for the three-dimensional flow field is known for an inclined disk with finite radius R subjected to streams of air. In their pioneering papers on parallel rotating disks subjected to air streams, Dennis et al. [1] and Booth and de Vere [2] recognized how important the effect of flow separation at the rim of the blunt disk is for flow and heat transfer behavior. In general, the angle of attack (i.e., the inclination) of the disk β represents a third major parameter in addition to rotational and translational Reynolds numbers Re_ω and Re_u. Furthermore, the disk thickness ratio d/R is also relevant for the occurrence of flow separation and reattachment of a turbulent boundary. It is therefore useful to organize the following discussion into two separate chapters: a discussion of the phenomena involved in a stationary disk (Chap. 5) followed by an extension to rotating disks (Chap. 6) based on the results for stationary disks.

5.1 Overview

For a stationary disk subjected to a stream of air, the mean Nusselt number

$$Nu_\mathrm{m} = \frac{h_\mathrm{m}R}{\lambda} = f_0(Re_u, \beta, d/R) \tag{5.1}$$

can generally be expressed as function f_0 of the translational Reynolds number Re_u, the angle of attack β, and the disk thickness ratio d/R for a fixed Prandtl number Pr. For a fixed disk thickness ratio, the mean Nusselt number Nu_m depends only on two major parameters, namely Re_u and β. Parameter β is the angle of attack (or incidence) and can be interpreted as a symmetry parameter. For a perpendicular disk, $\beta = 90°$, the flow is obviously axisymmetric, whereas in other cases the flow becomes fully three dimensional. For small deviations of β from the symmetric

© The Author(s) 2016
S. aus der Wiesche, C. Helcig, *Convective Heat Transfer From Rotating Disks Subjected To Streams Of Air*, SpringerBriefs in Applied Sciences and Technology, DOI 10.1007/978-3-319-20167-2_5

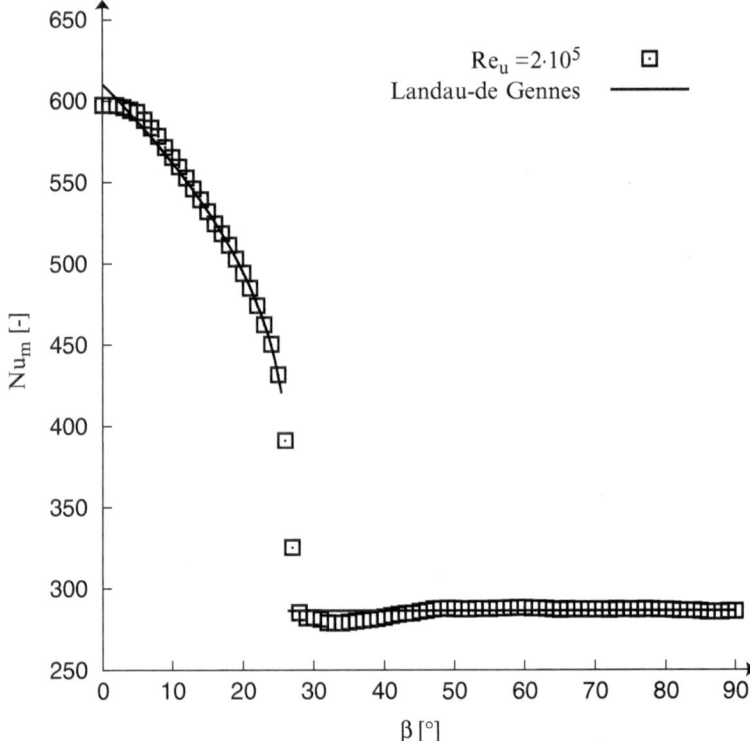

Fig. 5.1 Mean Nusselt number for a stationary disk with $d/R = 0.3$ subjected to a stream of air at $Re_u = 2 \times 10^5$ (data adapted from [3])

configuration, the flow might be still assumed as nearly axisymmetric, but a fully three-dimensional case is clearly achieved for a parallel disk, $\beta = 0°$, which also represents a limit case. Between both limit cases ($\beta = 0°$ and $\beta = 90°$), a symmetry breaking occurs that substantially affects the mean heat transfer behavior. This effect is illustrated in Fig. 5.1, where experimentally obtained mean Nusselt numbers Nu_m are plotted against the entire incidence range β for a fixed translational Reynolds number $Re_u = 2 \times 10^5$ for a disk with thickness ratio $d/R = 0.3$.

The experimental data shown in Fig. 5.1 were obtained with the disk apparatus described in detail in Chap. 3 and reference [3]. The turbulence factor of the wind tunnel was of order TF = 1.16, corresponding to a turbulence level of Tu = 0.2 %, which is a comparably low value for heat transfer experiments. Two remarkable observations can be made from the representative data of Fig. 5.1. First, the value of the mean heat transfer for the perpendicular disk can generally be taken for a large range of incidence, and it remains practically constant up to a transitional value β_{tr} of approximately $\beta_{tr} = 27° \pm 1°$ in the considered configuration. Second, at the transition value, a sharp increase of the mean heat transfer occurred. For incidence

$\beta < \beta_{tr}$, the convective heat transfer is much higher than for the stagnation flow and reaches its maximum in the case of a parallel blunt disk subjected to a stream of air.

In addition to the experimental data, Fig. 5.1 also plots the prediction of a theoretical model (namely the Landau-de Gennes model) for the transition. The background of the theoretical model will be discussed in more detail later. In the present case, the Landau-de Gennes model predicts the mean Nusselt number Nu_m as a function of the incidence angle β in accordance with

$$Nu_m = \frac{h_m R}{\lambda} = \begin{cases} Nu_{m,\beta=90°}(Re_u) & \text{for} \quad \beta \geq \beta_{tr} \\ C_1 + C_2\sqrt{1 - C_3(\beta - \beta_{tr})} & \text{for} \quad \beta < \beta_{tr} \end{cases}. \quad (5.2)$$

In correlation (5.2), the mean Nusselt number $Nu_{m,\beta=90°}$ for the perpendicular disk can be found using the methods discussed in Chap. 4. The fitting constants C_1, C_2, and C_3 depend on the Reynolds number Re_u, whereas the transition value β_{tr} depends mainly on the disk thickness ratio d/R, as discussed later in more detail. The limit value $\beta = 0°$ corresponds to a blunt disk subjected to a stream of air. The configuration of a parallel blunt flat *plate* has been investigated in detail [4–8]. The flow over a blunt plate is mainly characterized by flow separation at the leading edge and reattachment of a turbulent boundary layer. This were also observed for parallel blunt disks [1, 2] and are illustrated by the smoke visualization shown in Fig. 5.2. The separation bubble at the center plane has a length of order 4 to 5 times the disk thickness d. This result agrees well with the results for a blunt plate. The reattachment length is almost constant when the Reynolds number based on plate thickness and inflow velocity becomes greater than $u_\infty d/v = 700$ [6].

The mean heat transfer from a blunt plate or disk is comparably high due to the turbulent boundary layer and the strong velocity fluctuations in the vicinity of the separation bubble and the reattachment point. Based on this mechanism, the maximum mean heat transfer for a parallel disk at $\beta = 0$ in Fig. 5.1 can easily be explained. Its value can be obtained by means of correlations for a stationary disk subjected to a parallel stream of air.

5.2 Stationary Disk in a Parallel Stream of Air

The heat transfer from a rectangular flat surface in an airstream has been extensively studied [9]. Proceeding on the assumption of the boundary-layer theory for a flat plate, the following expressions for the local heat transfer coefficient have proven reliable:

$$h_x = 0.332 \lambda Pr^{1/3} \sqrt{\frac{u_\infty}{v}} \frac{1}{\sqrt{x}} \quad \text{(laminar flow)} \quad (5.3)$$

Fig. 5.2 Visualization of the flow past a blunt parallel disk ($d/R = 0.05$) with flow separation and reattachment of a turbulent boundary layer at $Re_u = 3.3 \times 10^4$

$$h_x = 0.0296\,\lambda\,Pr^{1/3}\left(\frac{u_\infty x}{\nu}\right)^{0.8}\frac{1}{x} \qquad \text{(fully turbulent flow)} \qquad (5.4)$$

In line with Dennis et al. [1], it is reasonable to employ the above local heat transfer coefficients for the flow over a circular parallel disk, too. The mean Nusselt number Nu_m for the entire circular disk can be found after integrating (5.3) and (5.4) over the entire disk surface. After some calculations [10], the following correlations result for $Pr = 0.71$:

$$Nu_m = \frac{h_m R}{\lambda} = 0.5968\,Re_u^{1/2} \qquad \text{(laminar flow)} \qquad (5.5)$$

$$Nu_m = \frac{h_m R}{\lambda} = 0.0307\,Re_u^{0.8} \qquad \text{(fully turbulent flow)} \qquad (5.6)$$

It should be noted that the originally stated values (see [1]) of the constants in the above correlations (5.5) and (5.6) were slightly erroneous; the corrected mathematical procedure can be found in reference [10]. Some experimentally obtained values for the mean Nusselt number Nu_m of a stationary parallel disk in a stream of air are plotted against the translational Reynolds number Re_u in Fig. 5.3. The measurements were performed with the disk apparatus and the wind tunnel described in Chap. 3 and in references [3] and [10] with disk thickness ratios of $d/R = 0.30$ and

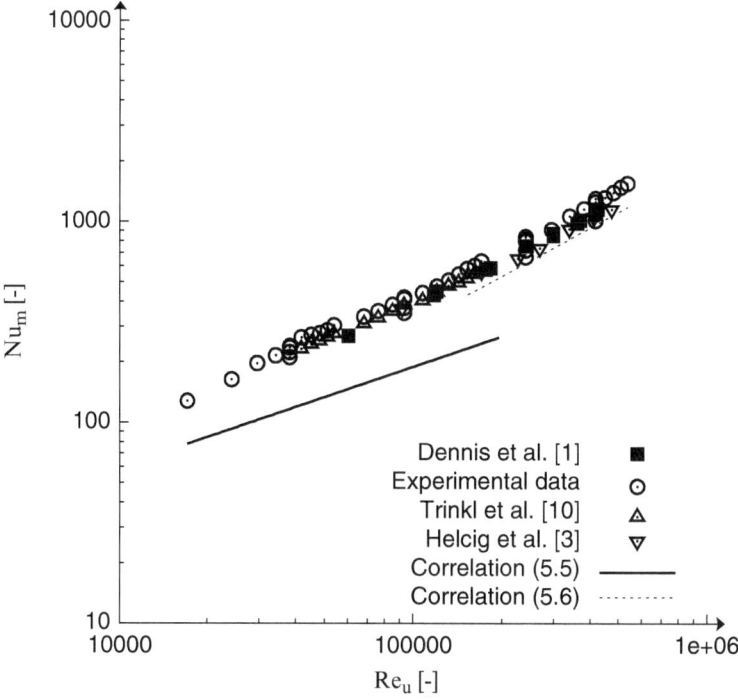

Fig. 5.3 Mean Nusselt number Nu_m against translational Reynolds number for a stationary disk subjected to a parallel stream of air

0.12. In addition to the experimental data, Fig. 5.3 also shows the theoretical correlations (5.5) and (5.6) and the early experimental results from Dennis et al. [1]. For a disk with a thickness ratio $d/R = 0.0978$, they found that the mean heat transfer from a stationary disk in a parallel stream of air agreed well with the empirical correlation

$$Nu_m = \frac{h_m R}{\lambda} = 0.036 Re_u^{0.8}, \tag{5.7}$$

which is noticeably higher than the theoretical prediction (5.6) or the recent experimental results. However, Dennis et al. [1] stated "that the freestream turbulence in the tunnel must have been high." Such an assumption could explain the higher level of their mean Nusselt numbers. On the other hand, the experimental results [10] obtained for a thin disk are in excellent agreement with the fully turbulent boundary-layer prediction (5.6) for a large range of Reynolds numbers. However, for a thicker disk with $d/R = 0.30$ [3], the mean heat transfer was slightly higher than predicted by (5.6). Here, the experimental data agreed better with an empirical correlation

$$Nu_{\mathrm{m}} = \frac{h_{\mathrm{m}}R}{\lambda} = 0.033\, Re_u^{0.8}. \tag{5.8}$$

The above effect of the disk thickness ratio d/R on the mean heat transfer can easily be explained on the basis of the flow separation illustrated in Fig. 5.2. The absolute length of the separation bubble increases with disk thickness d, as in the case of the flat blunt plate. Since stronger turbulent motion and heat transfer occur in the separation bubble and the reattachment point [11, 12], thicker parallel disks have slightly higher mean Nusselt numbers than the boundary-layer theory prediction (5.6) derived for a flat disk. Due to the flow separation at the leading edge of the blunt disk, the fully laminar flow regime is very difficult to achieve. Therefore the laminar boundary-layer theory prediction (5.5) derived for a purely laminar flow is mainly of academic interest for blunt disks with finite thickness.

5.3 Location of Stagnation Point and Bifurcation

The remarkably sharp transition (see Fig. 5.1) between two different mean heat transfer regimes at a certain transition angle β_{tr} can experimentally be observed only for comparably low inflow turbulence levels, and the value of the transition angle β_{tr} depends mainly on the disk thickness ratio d/R. Both observations and the empirical correlation (5.2) can be explained by means of the critical point and bifurcation theory introduced in Chap. 2.

For very low inflow turbulence, a subcritical bifurcation occurs for the flow regime at $\beta = \beta_{\mathrm{tr}}$. The control parameter Ψ is the incidence angle $90° - \beta$ for the present configuration since it governs the symmetry of the flow. A suitable order parameter Λ is given by the mean Nusselt number Nu_{m}. Over the range $\Psi < \Psi_{\mathrm{tr}}$ where the flow is stable, the order parameter $\Lambda = Nu_{\mathrm{m}}$ remains constant. The value for Nu_{m} is in the first order identical to the laminar stagnation flow result valid for the axisymmetric laminar flow over a perpendicular stationary disk. The order parameter remains constant over the wide range of $\Psi < \Psi_{\mathrm{tr}}$ (i.e., $\beta > \beta_{\mathrm{tr}}$) because the flow field remained topologically unchanged. The stagnation point is located on the disk surface for $90° \geq \beta > \beta_{\mathrm{tr}}$. At $\Psi = \Psi_{\mathrm{tr}}$ (i.e. $\beta = \beta_{\mathrm{tr}}$) the stagnation point is located at the edge of the blunt disk, and in the case of $\Psi > \Psi_{\mathrm{tr}}$ (i.e., $\beta < \beta_{\mathrm{tr}}$), the stagnation point is located at the rim side, and three-dimensional flow separation occurs at the leading edge. The occurrence of flow separation leads to a dramatic topological change of the flow field and breaks the symmetry. As a result, a finite jump occurs in the development of the order parameter Λ at $\Psi = \Psi_{\mathrm{tr}}$. The flow past a blunt disk is turbulent in the case of $\Psi > \Psi_{\mathrm{tr}}$ (i.e., $\beta < \beta_{\mathrm{tr}}$) due to the reattached turbulent boundary layer behind the separation bubble (see Sect. 5.2). This bifurcation is illustrated in Fig. 5.4, where the transition from a laminar stagnation flow to a turbulent flow with separation is shown schematically.

The location of the stagnation point depends on the incidence β and on the disk thickness ratio d/R. The transition occurs at $\beta = \beta_{\mathrm{tr}}$ when the stagnation point is

Fig. 5.4 Transition from a laminar stagnation flow (**a**) to a turbulent flow with separation and reattachment (**b**) and its relation to a subcritical bifurcation diagram

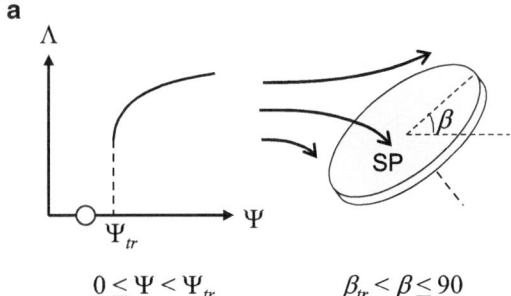

$$0 \leq \Psi < \Psi_{tr} \qquad\qquad \beta_{tr} < \beta \leq 90$$

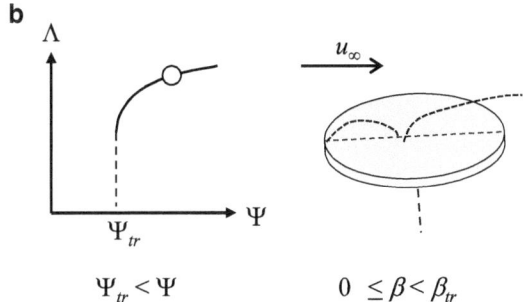

$$\Psi_{tr} < \Psi \qquad\qquad 0 \leq \beta < \beta_{tr}$$

located at the leading edge of the disk. It is therefore possible to calculate the transition value β_{tr} by means of computational fluid dynamics (CFD) simulations for the flow past an inclined blunt disk. Since only the location of the stagnation point is in question, and since a high Reynolds number level is involved, an inviscid flow field calculation is also applicable for the present purpose. The corresponding three-dimensional potential flow problem can be solved with powerful mathematical tools [13]. For an infinitely thin disk $(d=0)$, the relation between the distance y of the stagnation point from the disk center is given by the analytical expression

$$y = R \ \cos\beta. \tag{5.9}$$

In this limit case $(d=0)$, the stagnation point is always located on the disk surface for $\beta > 0$, and no bifurcation would occur for the disk with zero thickness and finite radius R. For a disk with finite thickness $(d > 0)$, the location of the stagnation point is mainly governed by the incidence β for high Reynolds number levels. Figure 5.5 plots the results for corresponding numerical simulations. The transition value β_{tr} is determined by the intersection with the $y/R - 1$ line in Fig. 5.5.

As a result, the numerical results shown in Fig. 5.5 predict a transition value of $\beta_{tr} = 26.6°$ for $d/R = 0.3$, which agrees extremely well with the experimental value

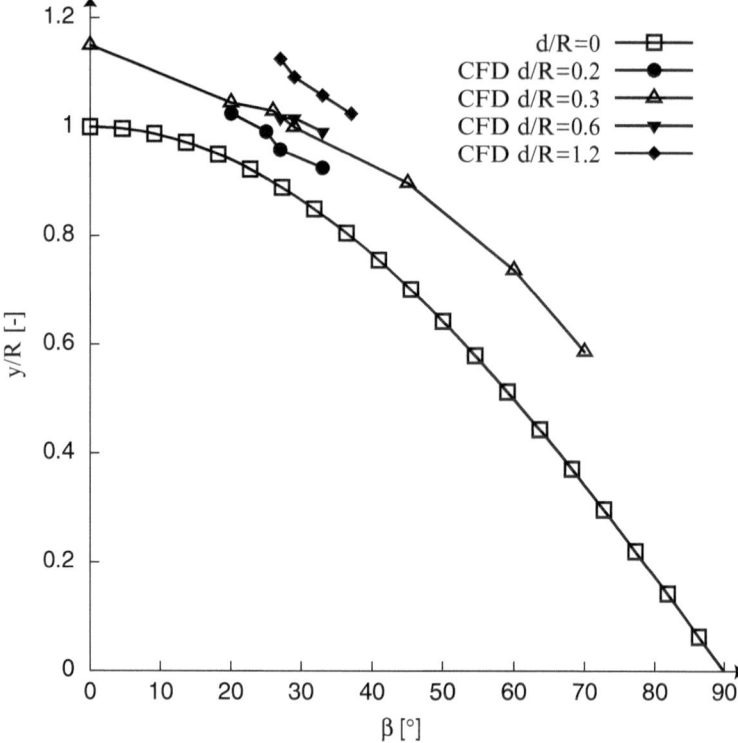

Fig. 5.5 Numerically obtained values for the location of the stagnation point

of $\beta_{tr} = 27° \pm 1°$ obtained for a Reynolds number of $Re_u = 2 \times 10^5$ (see Fig. 5.1). Based on numerical results (see examples plotted in Fig. 5.5), the simplified correlation

$$\beta_{tr} = 60.6° \left(\frac{d}{R}\right)^{1/2} - 21.9° \left(\frac{d}{R}\right) \tag{5.10}$$

for the transition value β_{tr} can be derived [14]. For a thin disk with ratio $d/R = 0.12$, the transition value is $\beta_{tr} = 18.4°$. This case was experimentally investigated, and the results of the mean Nusselt number Nu_m against Reynolds number Re_u are shown in Fig. 5.6 for some incidence angles β. The parallel disk ($\beta = 0°$) is characterized by a comparably high convective heat transfer caused by the turbulent boundary layer flow. For the thin disk ($d/R = 0.12$), the disks with incidence $20°$, $30°$, and $90°$ had practically the same mean heat transfer because the transition value was $\beta_{tr} = 18.4° < 20°$, which is smaller than the considered angles of attack in Fig. 5.6.

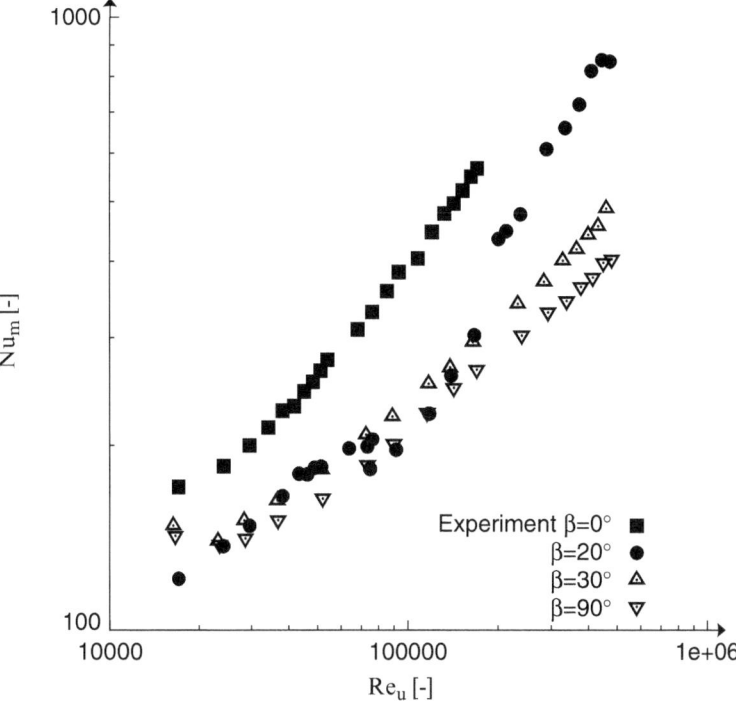

Fig. 5.6 Mean Nusselt number against Reynolds number Re_u of a thin inclined stationary disk ($d/R = 0.12$) for the same angles of attack above the transition value ($\beta_{tr} = 18.4°$)

5.4   Effect of Reynolds Number and Inflow Turbulence

For a sufficiently high Reynolds number Re_u, a turbulent boundary layer flow is established over the disk even where $\beta \geq \beta_{tr}$ because the laminar boundary layer due to the stagnation flow becomes unstable for sufficiently large flow paths. This situation is comparable with the well-known results for the flow past a flat plate [9], and it is illustrated in Fig. 5.7. In the case of the flow over a flat plate, the stability analysis shows that an unstable mode with wave length $\lambda_{pert} = 15\ \delta_1$ occurs if the Reynolds number $Re_{u,\delta} - u_\infty\ \delta_1/\nu$ defined by the boundary layer thickness δ_1 is of order $Re_{u,\delta} = 420$ (so-called point of indifference). Typically, the transition from laminar to turbulent flow occurs for a Reynolds number of order $Re_{u,y} = u_\infty\ y/\nu = 3 - 5 \times 10^5$ with y as a local coordinate measured from the leading edge. In cases with a very low inflow turbulence, it is possible to achieve transition values of order $Re_{u,y} = u_\infty\ y/\nu = 3 \times 10^6$. Transferred to the flow past an inclined disk ($\beta > \beta_{tr}$), the maximum flow length path y is given in the symmetry plane of the disk, as indicated in Fig. 5.7. For a sufficiently high local Reynolds number

Fig. 5.7 Transition to a
turbulent boundary layer
flow past a flat plate
(**a**) and an inclined disk
(**b**) for sufficiently high
Reynolds numbers

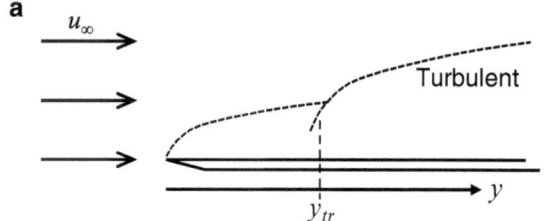

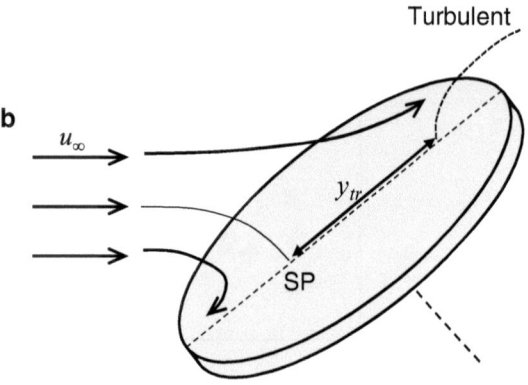

$Re_{u,y} = u_\infty y/v$, a transition to a turbulent boundary layer occurs. As a result, the mean convective heat transfer increases significantly. This effect was demonstrated by means of experimentally obtained values for the mean Nusselt number Nu_m for a stationary disk subjected to a stream of air, see Fig. 5.8, and it can be also recognized in Fig. 5.6 in case of the data for 20°. The disk thickness ratio was $d/R = 0.12$, yielding a transition value of $\beta_{tr} = 18.4°$ (see Sect. 5.3). The incidence angle of the disk was fixed to a value slightly higher than the transition value ($\beta = 20°$). For moderate Reynolds numbers Re_u, the mean convective heat transfer was fully in accordance with the predictions of the laminar boundary-layer theory, and the exponent of the heat transfer correlation $Nu_m = C (Re_u)^m$ was $m = 0.5$, as shown in Fig. 5.8. For a sufficiently high Reynolds number Re_u of order 10^5, the transition from laminar to turbulent boundary layer flow was observed as of a strong increase of the mean Nusselt number Nu_m. For $Re_u > 2 \times 10^5$, the fully turbulent heat transfer regime was established, as indicated by the exponent $m = 0.8$ for the heat transfer correlation.

For very low inflow turbulence level, it is possible to achieve fairly high values for the critical Reynolds number at which the transition to turbulent flow occurs. This possibility is demonstrated in Fig. 5.9, where a similar situation is shown for a thicker disk with thickness ratio $d/R = 0.3$ yielding a transition value of $\beta_{tr} = 26.6°$. For the low inflow turbulence level (with a turbulence factor of TF $= 1.16$ of the employed wind tunnel), the transition to turbulent boundary layer flow over the inclined disk occurred at a transition value of order $Re_u = 4 \times 10^5$.

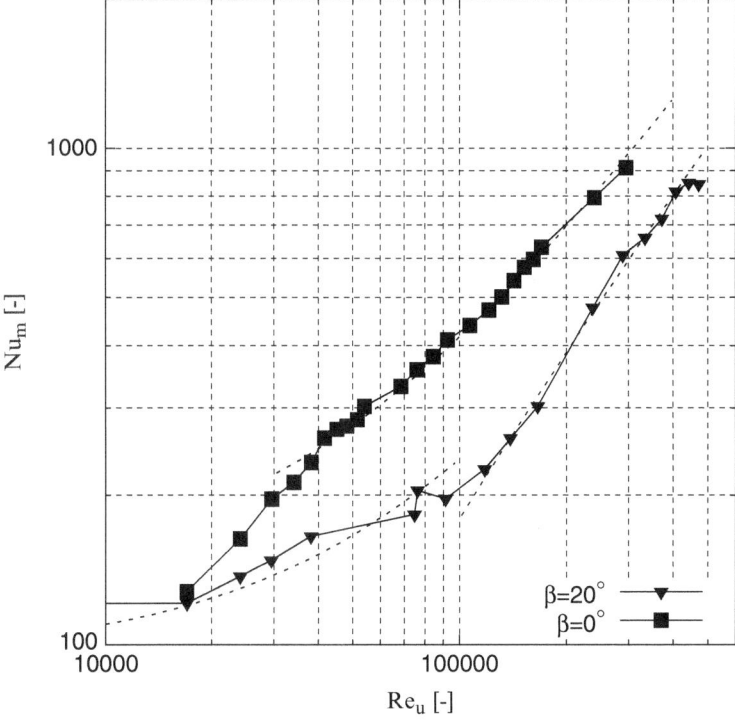

Fig. 5.8 Mean Nusselt number Nu_m against Reynolds number Re_u for an inclined stationary disk $(d/R = 0.12)$ at incidence $\beta = 20°$ $(\beta_{tr} = 18.4°)$

For higher inflow turbulence levels, the transition from a laminar to a turbulent boundary layer flow past the inclined stationary disk can occur for a fixed Reynolds number Re_u at a higher angle of attack $\beta > \beta_{tr}$ due to the mechanism discussed above. In such a case, no clear bifurcation phenomenon can be observed for the mean Nusselt number Nu_m as a function of incidence β because a turbulent boundary layer is not established by means of a topological change of the flow field.

The stagnation point is then still on the disk surface $(\beta > \beta_{tr})$, but the mean heat transfer is increased due to the turbulent boundary layer at the upper disk region. Since this turbulent flow field develops continuously from the laminar stagnation point flow, no jump or sharp increase for Nu_m occurs. This effect is illustrated in Fig. 5.10 where experimental data are plotted for a thick disk $(d/R = 0.3)$ at a fixed Reynolds number $Re_u = 2 \times 10^5$, see Fig. 5.10 for a wind tunnel with a higher turbulence factor TF = 1.3 corresponding to a turbulence level of order Tu = 0.4 %. For a very low inflow turbulence level, a sharp increase of the mean Nusselt number occur at the transition value $\beta_{tr} = 26.6°$ (see Fig. 5.1). Due to the comparably high value of the Reynolds number Re_u and due to the higher inflow turbulence level, the measured mean Nusselt number increases continuously for even higher angles of

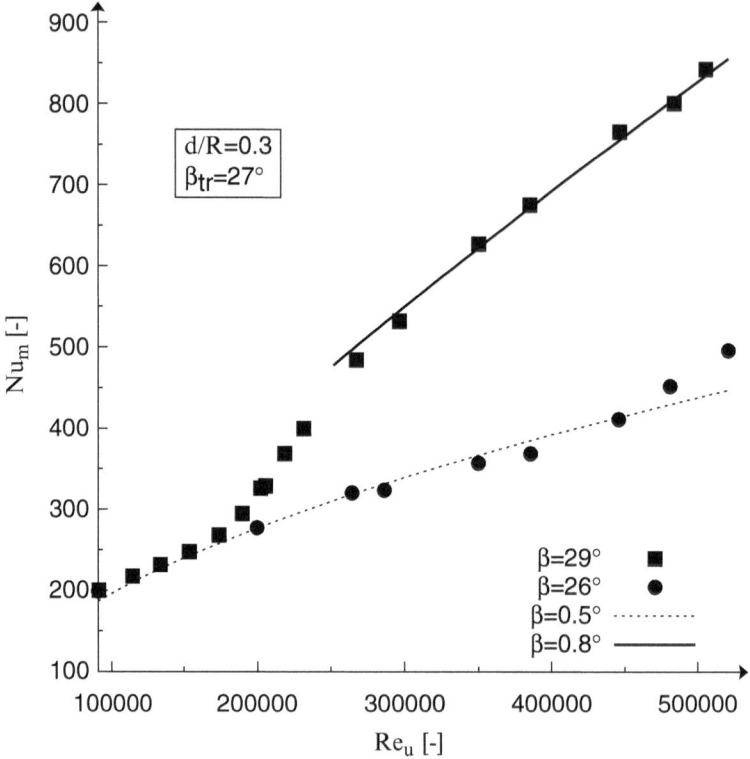

Fig. 5.9 Mean Nusselt number against Reynolds number for an inclined stationary disk ($d/R = 0.3$, $\beta = 29°$, $\beta_{tr} = 26.6°$) for a low inflow turbulence level (TF = 1.16)

attack $\beta_{tr} < \beta < \beta_{tr,lt}$ because a turbulent boundary layer was partially established on the disk surface for $\beta < \beta_{tr,lt} = 32°$. This example demonstrates how important the quality of the employed wind tunnel is when investigating boundary layer transitions. It is important to appreciate the limits of the idealized critical point and bifurcation theory in technical applications or experiments.

5.5   Phenomenological Model for the Transition at β_{tr}

It is possible to develop a simplified phenomenological model for the observed (idealized) transition in the vicinity of β_{tr} within the framework of open thermodynamic systems. Such a model is directly connected to the critical point and bifurcation theory introduced earlier, but it offers a valuable empirical correlation (5.2) without solving the complicated flow field equations. The main idea behind the approach is to interpret the present convective heat transfer situation as an open dynamic system as schematically illustrated in Fig. 5.11. Details about such a

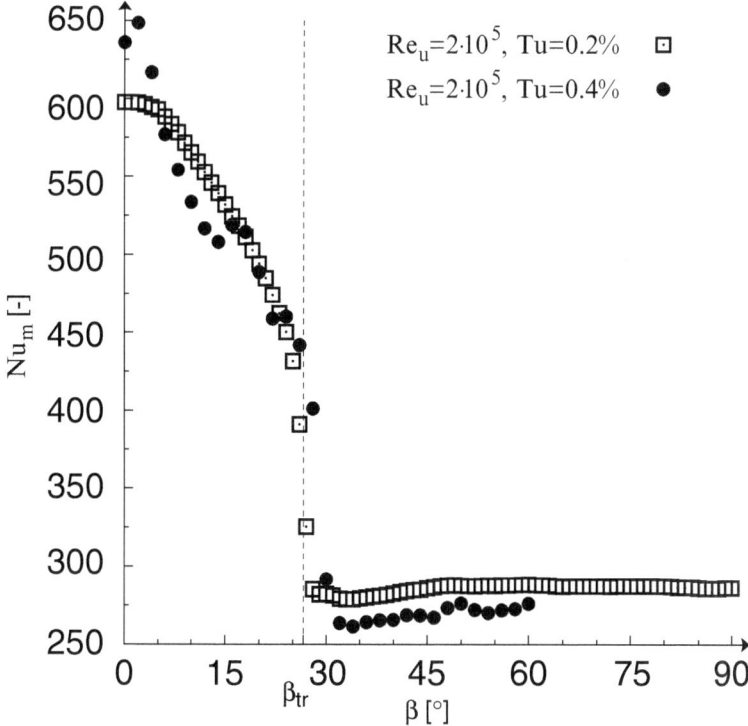

Fig. 5.10 Mean Nusselt number against Reynolds number for a stationary disk ($d/R = 0.3$) for a moderate inflow turbulence level (TF $= 1.3$)

Fig. 5.11 Heat transfer from a disk as a simple open dynamic system

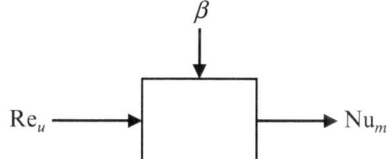

general approach can be found elsewhere [15]; in the following, only the main implications for the present heat transfer are discussed. The output of the system shown in Fig. 5.11 is the mean Nusselt number Nu_m. The input variable of the system are the Reynolds number Re_u and the incidence β. The latter is also related to the symmetry of the dynamics, and it is therefore interpreted as a parameter that governs system behavior.

In the general theory of dynamic systems, it is possible to describe the dynamics by means of a generalized potential ψ. In classic thermodynamics, such a potential is given, for instance, by the free energy or Gibbs potential [16], whereas in the case of open systems, the potentials are known as Landau, Landau-Ginzburg, or Landau-de Gennes potentials [17, 18].

The governing equation (or, in the present case, the convective heat transfer correlation) results after minimizing the potential

$$\delta\psi = 0. \tag{5.11}$$

The simplest expression for the corresponding potential as a function of the order and control parameters is given as a polynomial. In the present case, it is therefore suitable to consider

$$\psi = \psi_0 + \frac{1}{2}A(Re_u) \times (\beta - \beta_{tr})Nu_m^2 + \frac{1}{3}B(Re_u)Nu_m^3 + \frac{1}{4}C(Re_u)Nu_m^4 \tag{5.12}$$

as the corresponding potential in the lowest order. The value of the constant ψ_0 is arbitrary due to the minimization (5.11). The coefficients A, B, and C depend in principle on the Reynolds number Re_u as an input variable. Furthermore, the symmetry breaking at $\beta = \beta_{tr}$ leads to an additional dependency at least for the second term in expression (5.12). In the lowest order, the simplest approach for taking the symmetry breaking into account is given by the linear relationship assumed in (5.12). Minimizing (5.12) in accordance with (5.11) yields, after some minor manipulations, correlation (5.2) as stable values of Nu_m. Correlation (5.2) is mathematically the expression for the present subcritical bifurcation. The excellent agreement with experimental data is shown in Fig. 5.1. The general approach governs the qualitative behavior in the vicinity of the transition, but it is not possible to fix the values of the constants only on the basis of symmetry considerations.

References

1. Dennis RW, Newstead C, Ede AJ (1970) The heat transfer from a rotating disc in an air crossflow. In: Proceedings of 4th International Heat Transfer Conference, Versailles, 1970 (paper FC 7.1)
2. Booth GL, de Vere APC (1974) Convective heat transfer from a rotating disc in a transverse air stream. In: Proceedings of 5th International Heat Transfer Conference, Tokyo, 1974 (paper FC1.7)
3. Helcig C, aus der Wiesche S (2013) The effect of the incidence angle on the flow over a rotating disk subjected to forced air streams. In: Proceedings ASME Fluids Engineering Summer Meeting, Incline Village, NV (paper FEDSM2013-16360)
4. Ota T, Kon N (1974) Heat transfer in the separated and reattached flow on a blunt flat plate. ASME J Heat Transf 96:459–462
5. Ota T, Itasaka M (1976) A separated and reattached flow on a blunt flat pate. ASME J Fluids Eng 98:79–86
6. Yanaoka H, Yoshikawa H, Ota T (2003) Direct numerical simulation of turbulent separated flow and heat transfer over a blunt flat plate. ASME J Heat Transf 125:779–787
7. Kottke V, Blenke H, Schmidt KG (1977) Einfluß von Anströmprofil und Turbulenzintensität auf die Umströmung längsangeströmter Platten endlicher Dicke. Wärme- Stoffübertragung (Heat Mass Transfer) 10:159–174 (in German)

8. Kottke V, Blenke H, Schmidt KG (1977) Bestimmung des örtlichen und mittleren Stoffübergangs an längsangeströmter Platten endlicher Dicke mit Ablösen und Wiederanlegen der Strömung. Wärme- Stoffübertragung (Heat Mass Transfer) 10:159–174 (in German)

9. Schlichting H, Gersten K (1997) Grenzschicht-Theorie, 9th edn. Springer, Berlin

10. Trinkl CM, Bardas U, Weyck A, aus der Wiesche S (2011) Experimental study of the convective heat transfer from a rotating disc subjected to forced air streams. Int J Therm Sci 50:73–80

11. Nishiyama H, Ota T, Sato K (1988) Temperature fluctuations in a separated and reattached turbulent flow over a blunt flat plate. Heat Mass Transf 23:275–281

12. Helcig C, Uhkötter S, aus der Wiesche S (2014) Flow over a blunt disk with incidence: location of stagnation point and flow separation. In: Proceedings ASME Fluids Engineering Summer Meeting, Chicago, IL (paper FEDSM2014-21181)

13. Milne-Thomson LM (1996) Theoretical hydrodynamics. Dover Publications, New York

14. Helcig C, aus der Wiesche S (2014) Convective heat transfer from a free rotating disk subjected to a forced crossflow. In: Proceedings ASME Turbo Expo, Düsseldorf, Germany (paper GT2014-26223)

15. Haken H (1978) Synergetics, 2nd edn. Springer, Berlin

16. Falk G (1968) Thermodynamik. Heidelberger Taschenbücher, Springer, Heidelberg

17. Haug H (1997) Statistische Physik. Vieweg, Wiesbaden

18. Papon P, Leblond J, Meijer PHE (2006) The physics of phase transitions. Springer, Berlin

Chapter 6
Rotating Disk in Air Stream

The flow and heat transfer behavior for a stationary disk subjected to a uniform stream of air was discussed in Chap. 5. The various phenomena could be explained on the basis of a critical point and bifurcation theory and fundamental boundary layer considerations because mainly the translational Reynolds number Re_u and the incidence β govern the flow field. The extension to a rotating disk requires substantial efforts because a third major parameter, the rotational Reynolds number Re_ω, is now involved. Correspondingly, the flow behavior becomes much more complicated in comparison to the stationary disk. However, to a large extent it is possible to systematically discuss the new phenomena within the framework given by the limited case of a stationary disk. The rotational effects are then considered as perturbations. This approach is chosen in the present chapter as a start into the complex field of the flow over an inclined rotating disk.

6.1 Low Rotational Rates: Stagnation Flow Regime

In the axisymmetric configuration ($\beta = 90°$) of a rotating disk subjected to a perpendicular stream of air, it is possible to obtain analytical results by means of the self-similar solution approach. This approach was presented in detail in previous chapters. It is natural to assume that main relations such as the increase of the mean heat transfer due to rotation could be applied also to other incidence angles β within the stagnation flow regime ($\beta_{tr} < \beta \leq 90°$) because the flow field remains topologically unchanged. This assumption is supported by the observation (see Fig. 5.1) that, within the stagnation flow regime, the mean heat transfer from a stationary disk is practically identical to its axisymmetric expression.

In line with Mabuchi and coworkers [1–3], the increase of the mean Nusselt Number Nu_m from a rotating disk can be compared to the limit case of a stationary

© The Author(s) 2016
S. aus der Wiesche, C. Helcig, *Convective Heat Transfer From Rotating Disks Subjected To Streams Of Air*, SpringerBriefs in Applied Sciences and Technology, DOI 10.1007/978-3-319-20167-2_6

disk with heated radius R characterized by the mean Nusselt number $Nu_{m,\omega=0}$. Mabuchi and coworkers preferred the potential flow parameter

$$a = \frac{2}{\pi} \frac{u_\infty}{R} \tag{6.1}$$

for describing the forced stream of air with inflow velocity u_∞. The ratio between inflow and rotational rate is therefore given by the quantity

$$\frac{a}{\omega} = \frac{2}{\pi} \frac{Re_u}{Re_\omega}, \tag{6.2}$$

where the translational Reynolds number $Re_u = u_\infty R/\nu$ and the rotational Reynolds number $Re_\omega = \omega R^2/\nu$ are used. Interpreting the effects due to (low) rotational rates as a perturbation to the flow and heat transfer of a stationary disk, the dimensionless ratio Re_ω/Re_u can be chosen as an appropriate (small) perturbation parameter.

For *laminar* flow, the analytical treatment [1–3] yielded an expression

$$\frac{Nu_m}{Nu_{m,\omega=0}} = F(Re_\omega/Re_u) \times \left(1 + \frac{\pi^2}{4}\left(\frac{Re_\omega}{Re_u}\right)^2\right)^{1/4} \tag{6.3}$$

for the increase of the mean Nusselt number for a *perpendicular* disk in an air stream ($Pr = 0.7$). The values for the function F in (6.4) and hence the heat transfer augmentation can be calculated numerically; some values are listed in Table 6.1. Figure 6.1 shows the predicted heat transfer augmentation (called Theory, see (6.3)) and some experimental data. All data were obtained using the experimental apparatus described in Chap. 3 and references [4] and [5]. The translational and rotational Reynolds number levels remained within the laminar flow regime. Furthermore, the disk was always inclined within its stagnation flow regime (i.e., $\beta > \beta_{tr}$). Of note is that expression (6.3), which was initially derived only for an orthogonal disk ($\beta = 90°$), can be applied to other incidence values, too (within the stagnation flow regime). The heat transfer augmentation due to rotation is comparably weak in laminar flow within the stagnation flow regime, and only a moderate increase of the mean Nusselt number resulted, even in cases where the rotational Reynolds number is noticeably higher than the translational Reynolds number. Additionally, the laminar rotating boundary layer effect always contributes to the

Table 6.1 Values of the function F and the increase of heat transfer as predicted by correlation (6.3) in the case of a laminar flow regime for a perpendicular rotating disk ($Pr = 0.7$)	a/ω	Re_ω/Re_u	$F(Re_\omega/Re_u)$	$Nu_m/Nu_{m,\omega=0}$
	0	0	1.000	1.000
	0.1	0.318	0.957	1.012
	0.25	0.637	0.867	1.031
	0.5	1.273	0.729	1.091
	1	2.546	0.616	1.250
	2	6.366	0.538	1.705

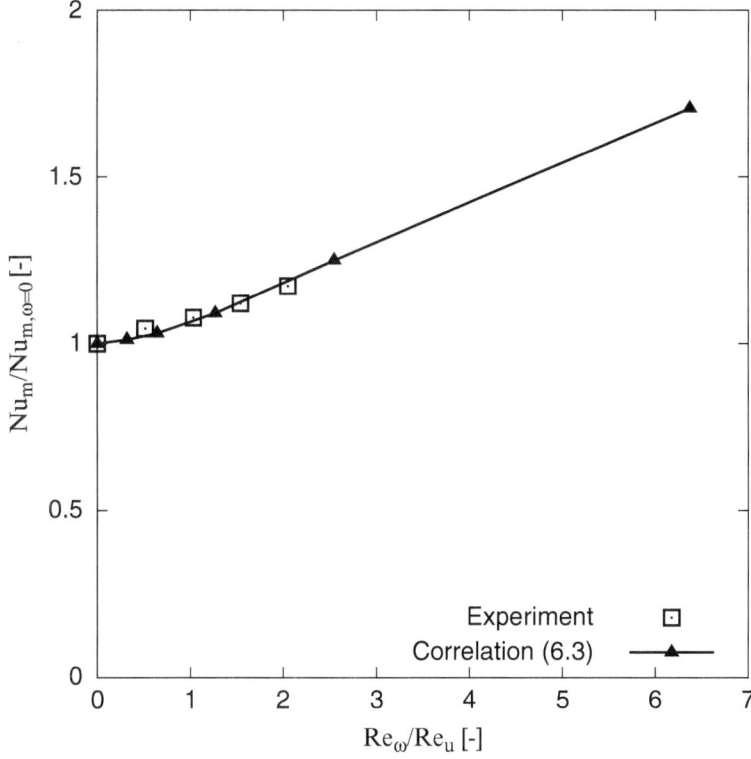

Fig. 6.1 Heat transfer augmentation due to rotation in the stagnation flow regime

mean heat transfer, and an augmentation occurs for $Re_\omega > 0$. The reason for this continuous increase is that the uniform jet velocity and the rotational speed interact and can be combined (see also the definition of a combined Reynolds number in Chap. 4). The topology of the stagnation flow field is not changed due to rotation, and a continuous development of the mean Nusselt number Nu_m as function of the Reynolds number ratio Re_ω/Re_u results.

6.2   High Rotational Rates: Stagnation Flow Regime

It is well known (see Chap. 4) that for a sufficiently high rotational Reynolds number Re_ω a transition from a laminar flow to a turbulent flow past a free rotating disk occurs. The corresponding transitional Reynolds number is of order $Re_{\omega,tr} = 2$ up to 3×10^5. The analytical treatment from Mabuchi and coworkers, i.e., correlation (6.3), is only applicable for Reynolds numbers below that transition value because a laminar flow regime is assumed over the entire disk. Even in cases where the ratio of the rotational and translational Reynolds numbers would be moderate,

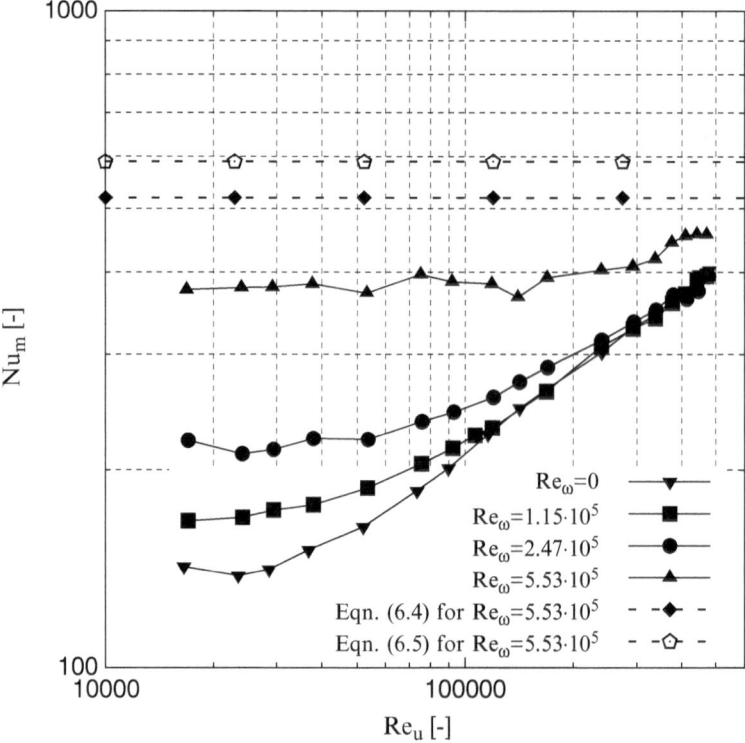

Fig. 6.2 Mean Nusselt number for a disk subjected to a perpendicular stream of air in case of a high rotational Reynolds number ($Re_\omega = 5.53 \times 10^5$)

the transition to turbulence caused by the three-dimensional instability for $Re_\omega > 3 \times 10^5$ represents a significant violation of the above assumptions. In such cases, the turbulent heat transfer due to the high rotational Reynolds number becomes dominant, and the contribution due to the laminar stagnation flow caused by the translational stream is of minor importance. This effect is shown in Fig. 6.2, where the measured mean Nusselt number Nu_m for an orthogonal disk ($\beta = 90°$) is plotted against the translational Reynolds number Re_u for a turbulent rotational Reynolds number $Re_\omega = 5.53 \times 10^5$. Furthermore, the figure also indicates the constant value of the mean Nusselt number for a free rotating disk without any forced stream. The experimental data agree well with that value, indicating that the convective heat transfer is practically governed by the turbulent rotating boundary-layer flow in that regime. Only in cases where the contribution of the laminar stagnation flow becomes equal or higher than the turbulent contribution is the mean heat transfer affected by the forced stream of air.

It should be noticed that the applied rotational Reynolds number $Re_\omega = 5.53 \times 10^5$ corresponds to a value that is noticeably above the transition value $Re_{\omega,\mathrm{tr}}$, but a large area part of a free rotating disk would still be within the laminar flow regime. Hence the value $Re_\omega = 5.53 \times 10^5$ corresponds to a transitional mean flow and transitional mean convective heat transfer regime. That regime can be described by an empirical correlation ($Pr = 0.71$)

$$Nu_\mathrm{m} = 0.4\frac{Re_{\omega,\mathrm{tr}}}{Re_\omega^{1/2}} + 0.015\left(Re_\omega^{0.8} - Re_{\omega,\mathrm{tr}}^{0.8}\sqrt{\frac{Re_{\omega,\mathrm{tr}}}{Re_\omega}}\right) \tag{6.4}$$

first proposed by Cobb and Saunders [6, 7]. The fully turbulent correlation ($Pr = 0.71$)

$$Nu_\mathrm{m} = 0.015\, Re_\omega^{0.8} \tag{6.5}$$

is typically achieved for large rotational Reynolds numbers of order $Re_\omega \geq 10^6$.

6.3   Low Rotational Rates: Parallel Disk

The flow past a blunt stationary disk is characterized by a separation bubble at the leading edge and the reattachment of a turbulent boundary layer. This result was discussed in detail in Chap. 5. As a result, the mean convective heat transfer is fairly high and can be correlated by means of a turbulent heat transfer correlation $Nu_\mathrm{m} = K \, (Re_u)^{0.8}$. For low rotational rates or small rotational Reynolds numbers Re_ω, it is reasonable to expect a weak influence of the additional rotation, and only a moderate heat transfer augmentation due to rotation should occur in this case. However, by means of numerical calculations including also Large-Eddy-Simulations (LES) [8–10] and experiments [5], it was found that the heat transfer behavior could be better described by a supercritical bifurcation (the so-called Landau model). This discovery means that the mean Nusselt number Nu_m for a parallel rotating disk subjected to an outer stream of air ($Pr = 0.71$) can be expressed by

$$\begin{aligned} Nu_\mathrm{m} &= Nu_{\mathrm{m},\omega=0} & \text{for}\quad Re_\omega/Re_u < R_{\mathrm{cr}}\\ Nu_\mathrm{m} &= Nu_{\mathrm{m},\omega-0} + C_{\mathrm{cr}}\left(\tfrac{Re_\omega}{Re_u} - R_{\mathrm{cr}}\right)^{m_{\mathrm{cr}}} & \text{for}\quad Re_\omega/Re_u \geq R_{\mathrm{cr}} \end{aligned} \tag{6.6}$$

with a so-called critical Reynolds number ratio R_{cr} and a critical exponent m_{cr} and a fitting constant C_{cr} for the increase of the mean Nusselt number Nu_m. The value $Nu_{\mathrm{m},\omega=0}$ is identical to the mean Nusselt of a stationary disk subjected to a parallel stream of air. Its value can be determined by means of the data provided in Chap. 5.

The resulting supercritical bifurcation is shown schematically in Fig. 6.3. The *control parameter* Ψ is the Reynolds number ratio Re_ω/Re_u for the present

Fig. 6.3 Mean heat transfer
as a function of the
Reynolds number ratio
corresponding to a
supercritical bifurcation

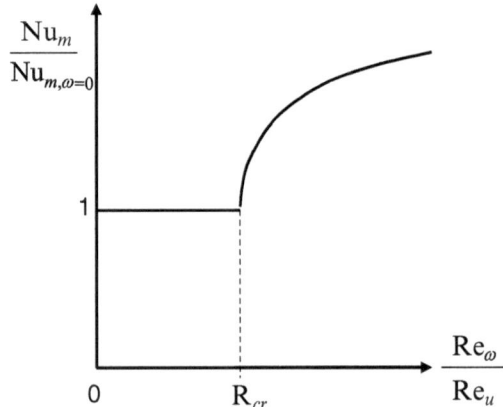

bifurcation. The *order parameter* Λ is given by the mean Nusselt number ratio $Nu_m/Nu_{m,\omega=0}$. Over the range $\Psi < \Psi_{tr} = R_{cr}$ where the flow past a disk is stable, the order parameter remains constant, $\Lambda = 1$. For $\Psi \geq \Psi_{tr} = R_{cr}$, a continuous increase of the order parameter Λ can be observed. This increase is characterized by a critical exponent m_{cr}. The bifurcation flow that replaces for $\Psi \geq \Psi_{tr} = R_{cr}$ the unstable flow can differ only infinitesimally from it, because the increase is continuously at a supercritical transition. This result contrasts strongly with the subcritical bifurcation discussed in Chap. 5 for the heat transfer from an inclined stationary disk. Because the bifurcation flow initially departs only infinitesimally from the unstable known flow (yielding $Nu_{m,\omega=0}$), the structural stability of the surface shear stress remains unaffected [11]. The results of the LES analysis [10] indicated a value for the critical exponent $m_{cr} = 0.5$, which is identical to the value predicted by the classic Landau model [12]. Furthermore, the critical ratio was found to be of order $R_{cr} = 1.4$ in the case of a parallel rotating free disk, and the constant was obtained to be $C_{cr} = 0.32$. Later, these numerical results were confirmed by experiments [5]. The supercritical bifurcation was found to be practically independent of the Reynolds number level. The Landau model can also be used to correlate the early experimental data obtained by Dennis et al. [7]. Furthermore, Booth and de Vere came to the same conclusion after their experiments [13], but they did not explicitly formulate a correlation.

The agreement with the ideal correlation (6.6) is typically not exact in real wind tunnel experiments due to noticeable inflow turbulence and measurement uncertainties, but the main feature can be identified even with comparable high wind tunnel turbulence. For smaller translational Reynolds numbers as considered in the LES studies [9, 10], the agreement with the predictions of the Landau model is quite remarkable. It should be mentioned that the validity range of the Landau model (6.6) is limited to certain values of the control parameter because, for higher values, the other contributions neglected in the derivation of the Landau model become relevant. The present results obtained for a parallel rotating disk indicated a maximal range of $\Psi < 5$ [10]. The agreement between experimental data and the

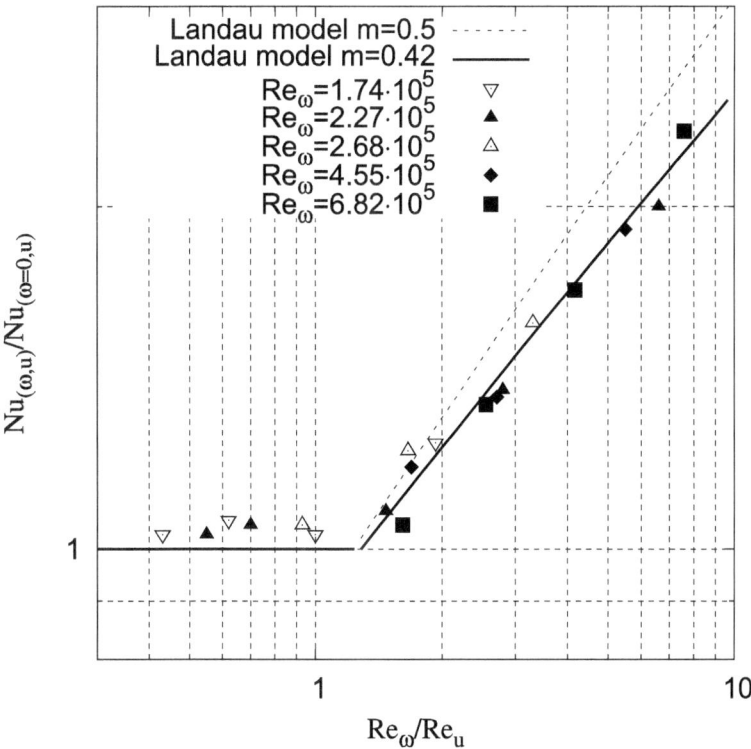

Fig. 6.4 Convective heat transfer augmentation due to rotation in the case of a parallel disk

predictions of the Landau model is shown in Fig. 6.4. The experimental values were obtained from a heated disk apparatus placed in the test section of a wind tunnel as described in Chap. 3.

The applicability of the Landau model was later independently confirmed by Latour et al. [13] for the convective heat transfer from a rotating disk mounted on a cylinder in a transverse air crossflow. For a disk mounted on a cylinder, different values for the constant and the exponent were stated by Latour et al. [14] (depending also on their definitions of the Reynolds numbers), but the main feature of a supercritical bifurcation was clearly observed, too.

The above experimental and numerical results clearly indicate the presence of a supercritical bifurcation at a certain value R_{cr} of the Reynolds number ratio Re_{ω}/Re_u, but the underlying mechanism or a derivation of its specific values has not yet been provided. This could be done on the basis of the critical point theory and a consideration of the corresponding potential flow. Often, the dynamics of a frictionless fluid is regarded as an academic subject and incapable of practical application, but in the present case, the potential flow provides a suitable approach for discussing the location of the stagnation points and the conditions of the observed transition. As an introductory step, it is helpful to review the well-known

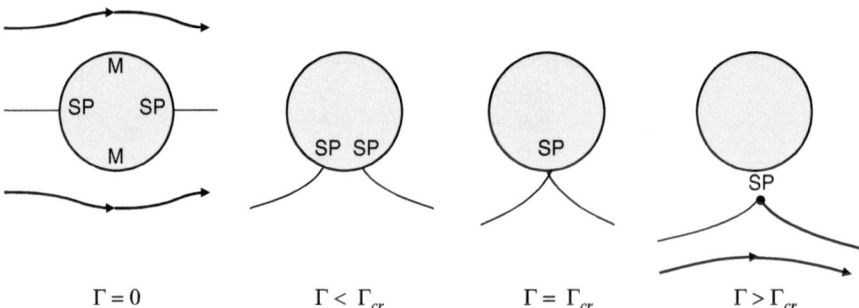

Fig. 6.5 Two-dimensional potential flow past a rotating cylinder and the location of the stagnation points

two-dimensional inviscid flow past a rotating cylinder with radius R and crossflow speed u_∞. Details can be found in textbooks [15], and the main features are illustrated in Fig. 6.5, where the location of the stagnation points on the cylinder surface is shown for three different values of circulation Γ or rotation rates ω.

For vanishing circulation $\Gamma = 0$ (i.e., $\omega = 0$), both stagnation points SP are located symmetrically on the cylinder surface, as shown in Fig. 6.5. The maximum velocity at point M is given by $u_M = 2\,u_\infty$ in the case of the two-dimensional potential flow past a stationary cylinder. With increasing circulation, the stagnation points move, as indicated by Fig. 6.5. For a certain value Γ_{cr}, there is only one stagnation point on the surface of the cylinder, corresponding to $\omega R = 2\,u_\infty = u_M$. For higher rotation, $\omega R > u_M$, the stagnation point moves into the fluid domain. The movement of the stagnation point corresponds to a supercritical bifurcation. Below the critical value, there are two stagnation points on the cylinder, but for supercritical values of the rotation rate, there is only one on the cylinder surface. The flow behavior in the case of a stagnation point within the fluid domain represents a substantial change for the mean heat transfer because the flow separation is affected. This change is, however, a smooth event, and hence a supercritical rather than a subcritical bifurcation results.

The above two-dimensional potential flow past a rotating cylinder is treated in many textbooks, but it is a straightforward matter to extend it and consider three-dimensional cases. Here, it is mathematically convenient to consider in a next step the flow past an ellipsoid described by the general expression

$$\frac{x^2}{a^2} + \frac{y^2}{b^2} + \frac{z^2}{c^2} = 1. \tag{6.7}$$

If the translatory motion of the ellipsoid in a *resting* fluid is in the direction of the x-axis, the velocity potential is given by

$$\varphi = \frac{abcu_\infty x}{2 - \alpha_0} \int_\lambda^\infty \frac{d\lambda}{(a^2 + \lambda)^{3/2}(b^2 + \lambda)^{1/2}(c^2 + \lambda)^{1/2}}, \tag{6.8}$$

and on the ellipsoid

$$\varphi = \frac{\alpha_0 u_\infty x}{2 - \alpha_0},$$ (6.9)

with

$$\alpha_0 = abc \int_0^\infty \frac{d\lambda}{(a^2 + \lambda)^{3/2} (b^2 + \lambda)^{1/2} (c^2 + \lambda)^{1/2}}.$$ (6.10)

The velocity u_M at point M, see Fig. 6.6, is given by the analytical expression

$$u_M = -\frac{\partial \varphi}{\partial x} = -\frac{\alpha_0}{2 - \alpha_0} u_\infty.$$ (6.11)

In the case of the uniform flow past a *stationary* ellipsoid, the above result (6.11) can easily be transformed into the new inertial system by means of $-u_\infty$. The flow past an ellipsoid is sketched in Fig. 6.6. The general expression (6.10) can be evaluated using elliptic integrals [16]. It is hence possible to obtain an exact expression for the maximum velocity u_M and the constant α_0 as a function of the shape of the ellipsoid governed by the parameters a, b, and c. Some values are listed in Table 6.2. For simplicity, the ellipsoids are limited to spheroids (i.e., $a = b$), and the only shape parameter is hence the ratio c/a. In that case, the elliptic integral, see expression (6.10), reduces to elementary functions. The special case $a = b = c$ is the sphere; c tending to infinity represents the flow past an infinite cylinder. For the flow past a parallel disk, the oblate spheroids with $c/a < 1$ represent suitable idealized configurations. It should be noticed that the values listed in Table 6.2 can be expressed by exact elementary expressions. The special case of a sphere can be elementarily obtained, and the result $u_M = 3u_\infty/2$ can be found in textbooks, too. The bifurcation occurs at $\omega R = u_M$ ($R = a = b$). This leads to the prediction of a critical Reynolds number ratio R_{cr} for the parallel disk mainly in accordance with the values of the column u_M/u_∞ listed in Table 6.2. An inspection of Table 6.2 indicates that the experimentally observed values for $R_{cr} = 1.4 \pm 0.1$ for blunt disks with thickness ratios d/R of order 0.12 up to 0.3 are compatible with the predictions of the potential flow approach. The deviation can be easily explained on the basis that a blunt disk is covered only roughly by the assumption of a smooth prolate spheroid.

Fig. 6.6 Flow past an ellipsoid

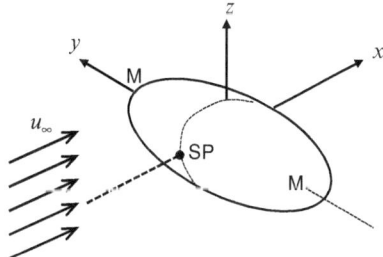

Table 6.2 Values for the velocity u_M for some ellipsoidal shapes ($a = b$, c variable)

c/a	α_0	u_M/u_∞	Shape
1	2/3	3/2	Sphere
0.6	0.52417	1.35517	Prolate spheroid
0.5	0.47280	1.30959	Prolate spheroid
0.4	0.41185	1.25933	Prolate spheroid
0.2	0.24952	1.14254	Prolate spheroid
0.1	0.13920	1.07481	Prolate spheroid
2	0.82643	1.70420	Oblate spheroid
4	0.92459	1.85976	Oblate spheroid
10	0.97971	1.96023	Oblate spheroid
∞	1	2	Cylinder

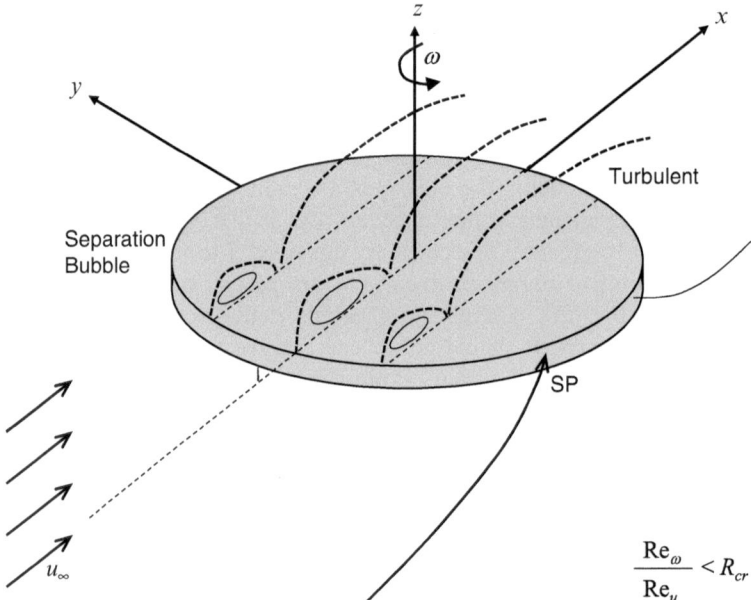

Fig. 6.7 Three-dimensional flow past a parallel rotating blunt disk and separation bubbles

An exact treatment of the present flow problem would require considering three-dimensional boundary layers and separation phenomena in the vicinity of a rotating surface, something that is beyond the present analytical scope. However, the above idealized approach is well able to explain the major mechanism for the supercritical bifurcation, and it predicts values that are in reasonable agreement with the observations for a parallel rotating disk. The schematics of the resulting three-dimensional flow past a parallel rotating blunt disk including separation bubbles at the leading edge are illustrated in Fig. 6.7.

6.4 High Rotational Rates: Exact Solution for a Parallel Disk

The problem of boundary layers in rotating flows is substantially complicated by the added effect of translating boundaries that destroy the axisymmetric character of the basic rotating flow. However, for an infinite rotating plane subjected to a parallel uniform stream, an exact solution of the Navier–Stokes equation for an incompressible fluid can be obtained in line with Rott and Lewellen [17]. Their solution offers a good starting point for discussing the flow and convective heat transfer from a parallel rotating disk at very high rotational Reynolds numbers (or $Re_\omega/Re_u \gg 1$). Although the turbulent flow deviates significantly from Rott and Lewellen's laminar analytical boundary layer solution due to the finite radius R of the disk, the main features of the observed flow and heat transfer behavior are still covered. Their treatment is fully applicable to the high rotational rate regime because the main assumption of the following analytical treatment is the practically constant boundary-layer thickness. This assumption is fulfilled for laminar flow over a free rotating disk, see Chap. 4.

The coordinate system used in the following analysis is sketched in Fig. 6.8. The rotating plane is defined by $z = 0$. The uniform stream is directed in the x-axis and, far away from the plane, has the constant velocity u_∞. The use of Cartesian coordinates means that the resulting three-dimensional velocity field can be expressed by a self-similar approach (see Chap. 4)

$$u_x = \omega\left(xF' - yG\right) + u_\infty H_1 \tag{6.12}$$

$$u_y = \omega\left(xG - yF'\right) + u_\infty H_2 \tag{6.13}$$

$$u_z = -2\sqrt{\omega v}F \tag{6.14}$$

Fig. 6.8 Uniform flow past a rotating plane

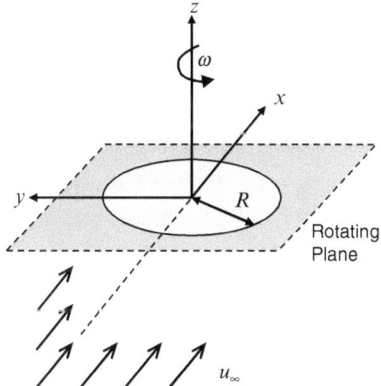

where F, G, H_1, and H_2 are functions of the variable

$$\zeta = z\sqrt{\omega/\nu}. \tag{6.15}$$

The classic von Karman flow results for $u_\infty = 0$. The above set of equations satisfies the continuity equation for an incompressible fluid, and inserting it into the stationary Navier–Stokes equations yields (after some rearranging and collecting of terms)

$$F''' + 2FF'' - F'^2 + G^2 - G(\infty) = 0 \tag{6.16}$$

$$G'' + 2FG' - 2F'G = 0 \tag{6.17}$$

$$H_1'' + 2FH_1' - F'H_1 + GH_2 = 0 \tag{6.18}$$

$$H_2'' + 2FH_2' - F'H_2 - GH_1 = 0 \tag{6.19}$$

The above set of ordinary differential equations (6.16)–(6.19) represents an exact solution of the Navier–Stokes equations. The pressure field is calculated after solving the velocity functions F, G, H_1, and H_2 (as in the case of the classic von Karman solution). The term $G(\infty)$ in (6.16) stems from the pressure gradient. The boundary conditions are:

$$\zeta = z\sqrt{\omega/\nu} = 0: \quad F(0) = F'(0) = 0, \quad G(0) = 1, \quad H_1(0) = 0, \quad H_2(0) = 0 \tag{6.20}$$

$$\zeta \to \infty: \quad F' \to 0, \quad G \to 0, \quad H_1 \to 1, \quad H_2 \to 0 \tag{6.21}$$

The above set of ordinary differential equations (6.16)–(6.19) and the boundary conditions (6.20) and (6.21) can be solved numerically. Corresponding values for H_1 and H_2 as functions of ζ can be found in [17]; the self-similar boundary layer profiles for the velocity are plotted in Fig. 6.9. The boundary-layer profiles related to the uniform stream, H_1 and H_2, are substantially thicker than the rotating boundary layer described by F' and G. Thus for a fluid with $Pr = 0.7$, the convective heat transfer would be primarily governed by the rotational boundary layer. The mean Nusselt number should then be nearly identical with the value obtained for a free rotating disk. This behavior is a consequence of the assumption of a constant boundary layer thickness. As discussed by Rott and Lewellen [17], the "secondary" translational flow dominates the "primary" rotational flow only within an "eye" of radius $R^* = u_\infty/\omega$. If R^* is much less than the radius R of a finite rotating disk, the secondary flow calculated under the above assumption represents a valid approximation, and the flow field features for the present exact solution should be applicable to the parallel rotating disk.

Another interesting feature of the present exact solution is the location of the vanishing resultant radial and circumferential shear. The loci of vanishing resultant radial and circumferential shear are circles, and they intersect each other at two points, see Fig. 6.10. The intersection at point N leads to a vanishing shear, but the

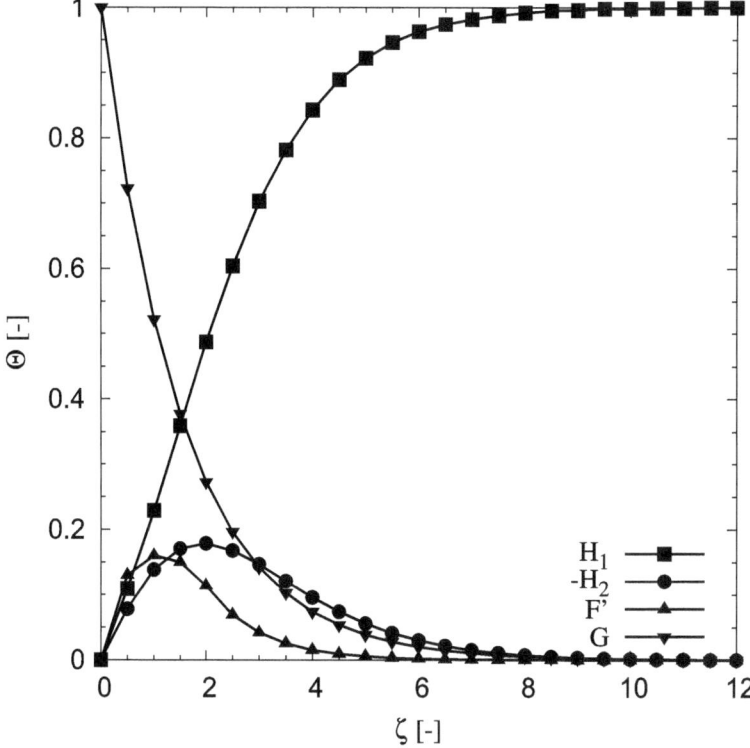

Fig. 6.9 Boundary layer profiles for a stream over a rotating plane

Fig. 6.10 Loci of vanishing resultant radial and circumferential shear (adapted from [17])

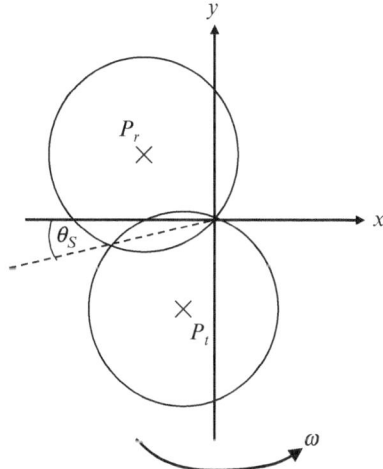

resultant velocity at $z = 0$ and x, $y > 0$ is not zero due to rotation. Hence, flow separation does not occur at this point. The other intersection at the origin of the coordinate system does not lead to a vanishing shear because the circumferential shear is singular at the origin. The coordinates of the centers of the circles for vanishing shear are given as follows [17]: P_r ($-0.208\,R^*$, $0.160\,R^*$) and P_t (-0.132 R^*, $-0.172\,R^*$), see Fig. 6.10. Since the flow separation criterion (see Chap. 2) is not met on the rotating plane, it can be concluded that flow separation for rotating parallel disks is governed by the leading edges or by singular points linked to the edge walls and not to the disk surface. This conclusion agrees well with the observations for finite disk systems. Furthermore, the intersection points of vanishing shear flow define an alignment angle $\theta_s = 12.8°$, see Fig. 6.10. Its existence can be interpreted as a "rotation" of the flow and heat transfer distribution from the viewpoint of an absolute frame of reference. Such an effect was indeed observed in the numerically calculated local Nusselt number distribution for a very thin rotating disk subjected to a parallel stream of air [10]. An illustration of such a distribution is shown in Fig. 6.11.

Details about the underlying LES simulations can be found elsewhere [10]. Due to the separation bubble at the leading edge, a half-moon-shaped area of a high local Nusselt number level can be clearly seen in Fig. 6.11. This half-moon is rotated in accordance with the angle θ_s. Numerically, the results shown in Fig. 6.11 indicated a rotating angle of $\theta_s = 12.5 \pm 0.5°$, which agrees excellently with the result based on the analytical treatment. It is quite remarkable that the value of the angle is not affected by the ratio of the Reynolds numbers. As in the case of the analytical treatment, the value of θ_s is universal in the sense that it is independent of Re_ω or Re_u (in case of $\omega > 0$ and $u_\infty > 0$).

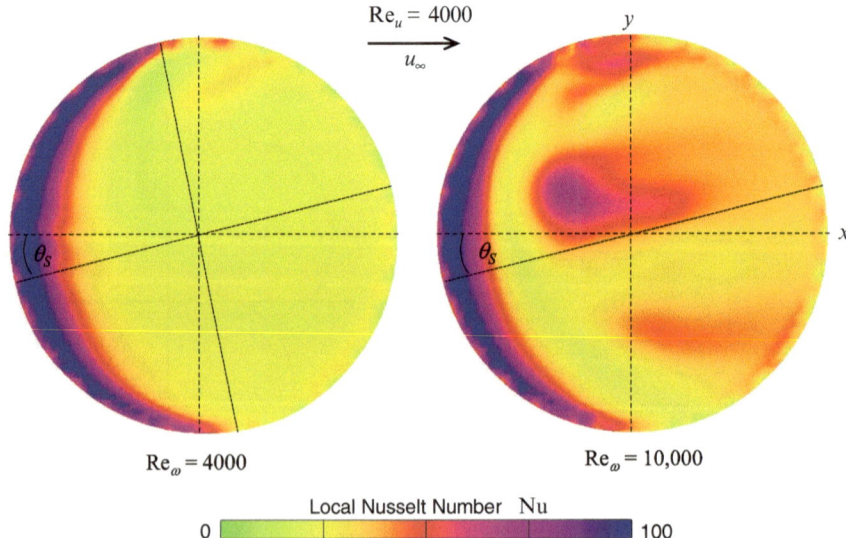

Fig. 6.11 Local Nusselt number distribution at the surface of a parallel rotating disk (adapted from [10])

6.5 Arbitrary Rotational Rates: Parallel Disk

Experimental data obtained with a combination of rotation and crossflow for a parallel disk were published as a plot of the resulting mean Nusselt number Nu_m against the translational Reynolds number Re_u for a series of rotational Reynolds number Re_ω in 1970 by Dennis et al. [7] and later in 2011 by Trinkl et al. [5]. The results of the two studies are closely aligned, and some representative data are shown in Fig. 6.12. At higher translational Reynolds numbers, the lines tend to come together and agree with the turbulent heat transfer correlation obtained for a stationary disk subjected to a parallel stream of air, indicating that the rotational flow has only a small effect on the heat transfer, and the crossflow dominates. At the other range, $Re_\omega \gg Re_u$, the lines tend to the horizontal, indicating that then the Nusselt number Nu_m is becoming independent of the translational Reynolds number. In that limit case, the heat transfer correlation obtained for a rotating disk in still air becomes applicable.

The graph may be presented the other way round, as is shown in Fig. 6.13, where the mean Nusselt number Nu_m is plotted against the rotational Reynolds number

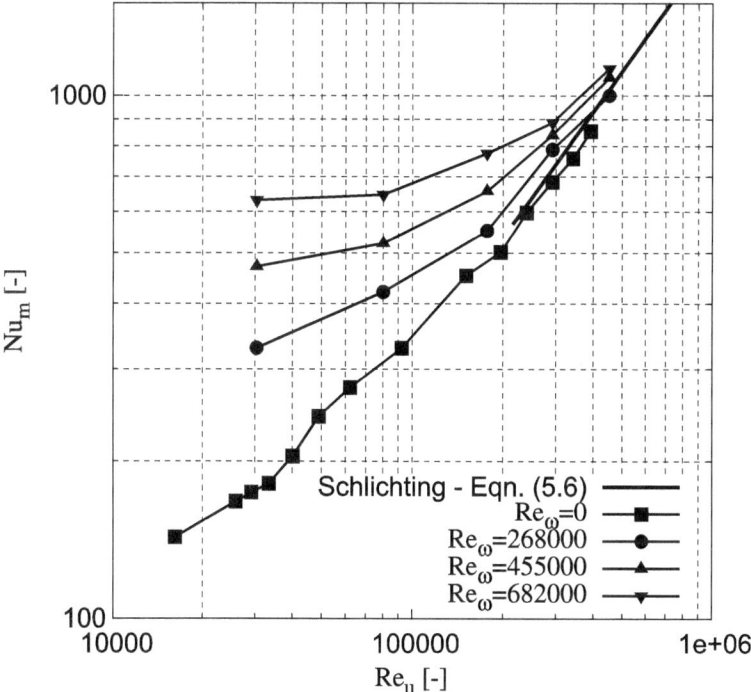

Fig. 6.12 Mean Nusselt number Nu_m against translational Reynolds number Re_u for a rotating disk subjected to a parallel air stream (adapted from [5])

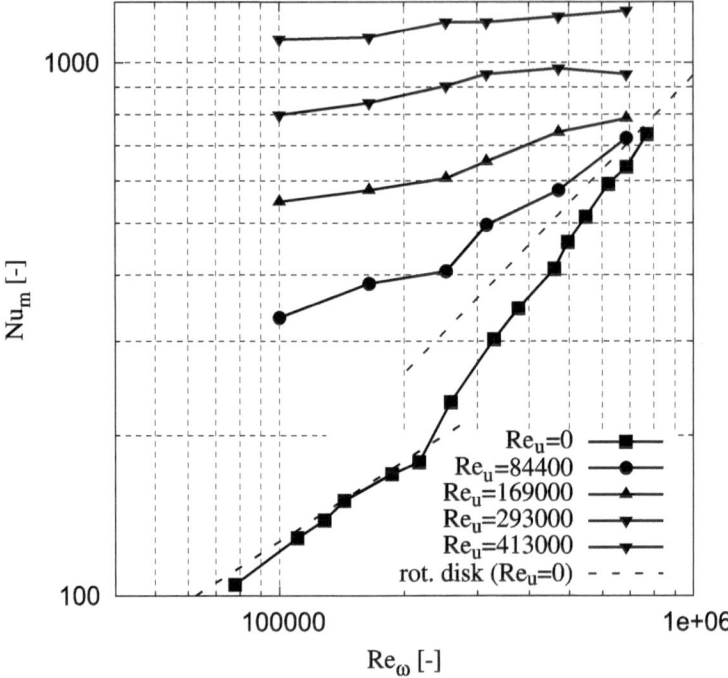

Fig. 6.13 Mean Nusselt number Nu_m against rotational Reynolds number Re_ω for a rotating disk subjected to a parallel air stream (adapted from [5])

Re_ω for a series of translational Reynolds numbers Re_u. A similar picture results, but Figs. 6.12 and 6.13 show that the influence of the rotational Reynolds number.

The stream of air leads to a turbulent flow due to flow separation at the leading edge even at comparably low translational Reynolds numbers. This feature was observed in early investigations [7, 13], too.

6.6 Inclined Rotating Disk

Flow and convective heat transfer from an inclined rotating disk (with an angle of attack β to the incident flow) were studied only very recently. The first experimental data obtained for some values of β were published in the literature in 2010 and 2011 [5, 18]. The first comprehensive experimental heat transfer study covering the whole range $0° < \beta < 90°$ with a high resolution $\Delta\beta = 0.5°$ was conducted by Helcig et al. in 2013 and 2014 [4, 19, 20]. In that study, the mean heat transfer from an inclined stationary disk and from an inclined rotating disk was measured. The case of an inclined stationary disk was discussed in detail in Chap. 5. An interesting feature was the observation of a subcritical bifurcation at $\beta = \beta_{tr}$. For higher angles of attack, the stagnation flow regime governed the heat transfer for the

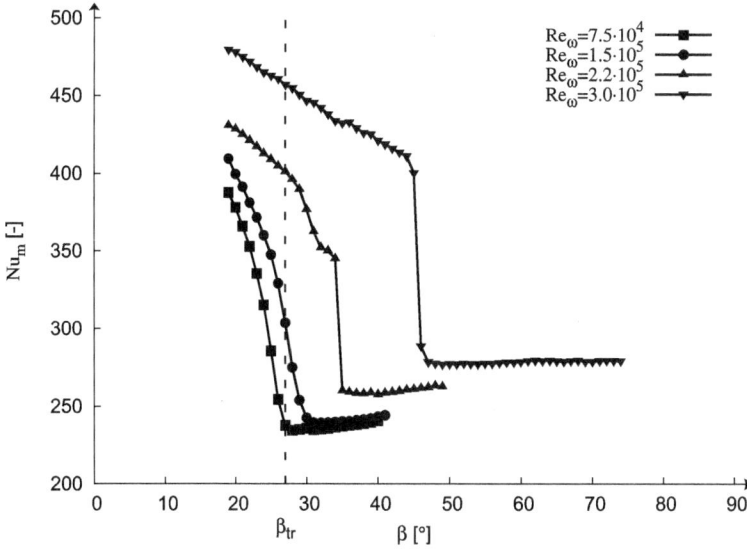

Fig. 6.14 Mean Nusselt number Nu_m against angle of attack β for a series of rotational Reynolds numbers Re_ω ($d/R = 0.3$, $Re_u = 1.5 \times 10^5$)

stationary disk, whereas for small angles of attack, the flow separation at the leading edge yielded comparably high heat transfer rates. For an inclined rotating disk, a new bifurcation was observed at a value $\beta_{tr,\omega} > \beta_{tr}$. To illustrate the rather complex heat transfer behavior connected to the translational and rotational Reynolds numbers and angle of attack for a disk with finite thickness, the mean Nusselt number Nu_m is plotted against the angle of attack β in Fig. 6.14 for a fixed translational Reynolds number $Re_u = 1.5 \times 10^5$ and a series of rotational Reynolds numbers Re_ω. The mean Nusselt number values were obtained using an electrically heated disk apparatus with a thickness ratio $d/R = 0.3$, yielding a transition value of $\beta_{tr} = 26.6°$ (see Chap. 5). The inflow turbulence level during the experiments was of order $Tu \leq 0.2$ %. That comparably low inflow turbulence level enabled the observation of a fairly sharp transition at $\beta = \beta_{tr,\omega} > \beta_{tr}$ for sufficiently high rotational Reynolds numbers (see the lines for $Re_\omega = 2.2 \times 10^5$ and 3×10^5). Its value depended on the rotation rate.

Obviously, the mean heat transfer for a disk at high incidence, $\beta > \beta_{tr,\omega}$, is practically governed by the stagnation flow, as in the case of a stationary disk. Then, the value for the mean Nusselt number Nu_m can be obtained by the correlations presented earlier. For small rotational rates, the heat transfer augmentation due to rotation is weak, see Sect. 6.1. For higher rotational rates, the Nusselt number increases due to the combined effect of rotation and stagnation flow, but this increase is moderate within the stagnation flow regime for $\beta > \beta_{tr,\omega}$. For these rates, the stagnation point is then located on the disk surface as schematically shown in Fig. 6.15a. Flow separation does not occur for sufficiently low inflow turbulence. For small angles of attack, $\beta < \beta_{tr}$, the stagnation point is located on the edge wall of the blunt disk, and flow separation and reattachment of a turbulent boundary layer occur, Fig. 6.15b. As a consequence, the mean Nusselt number has comparably high values and tends at $\beta \to 0$ towards to the turbulent correlation

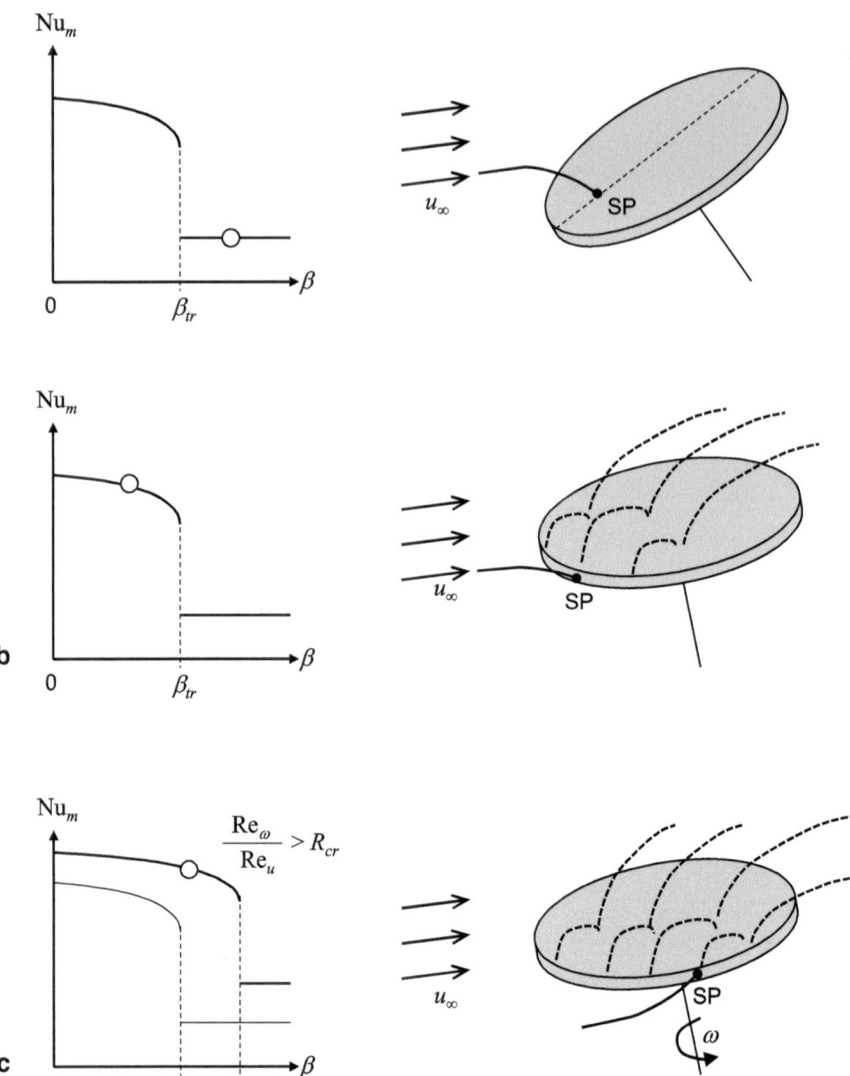

Fig. 6.15 Stagnation point location and flow field behavior for an inclined rotating disk subjected to a forced stream: (**a**) Stagnation flow regime. (**b**) Flow separation at small rotation rates and small incidence. (**c**) Flow separation at high rotation rates

obtained for the turbulent flow past a parallel disk. In this regime, the ratio of the Reynolds number Re_ω/Re_u is relevant for the heat transfer augmentation due to rotation, see Sect. 6.3. For the limit case $\beta = 0$, a supercritical bifurcation results for the control parameter $\Psi = Re_\omega/Re_u$. For sufficiently high rotational rates, $Re_\omega/Re_u > R_{cr}$, the mean heat transfer, i.e., the order parameter increases. For non-parallel disks with moderate incidence, $0 < \beta < \beta_{tr}$, the combination of inflow

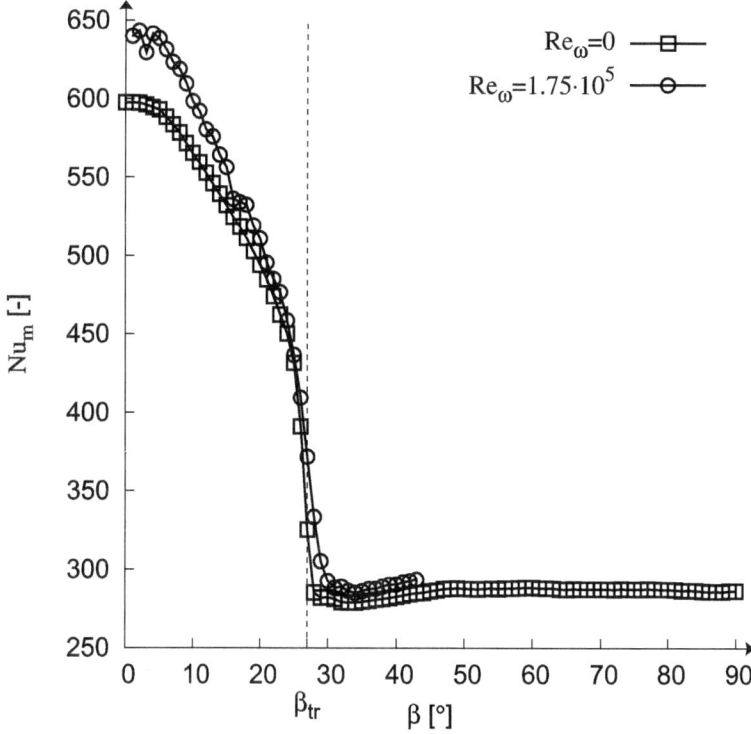

Fig. 6.16 Mean Nusselt number Nu_m against angle of incidence β in the vicinity of the subcritical bifurcation for $Re_u = 2 \times 10^5$ and $d/R = 0.3$ (data adapted from [4])

and rotation leads to a combined heat transfer, and the mean Nusselt number is a function of both Reynolds numbers, even in the cases $Re_\omega/Re_u \leq R_\text{cr}$. It can be furthermore observed that the transition at $\beta = \beta_\text{tr}$ smoothes out for $Re_\omega > 0$. This smoothing is shown in Fig. 6.16, where some experimental data are plotted in the vicinity of the subcritical bifurcation for a stationary disk and a rotating disk.

The transition value β_tr is not substantially affected due to the low rotation rate, as indicated by means of the experimental data shown in Fig. 6.16. Furthermore, the slight heat transfer augmentation due to rotation becomes noticeable within the stagnation flow regime for $\beta > \beta_\text{tr}$ in Fig. 6.16. For $0 < \beta < \beta_\text{tr}$, the rotation resulted also in a slightly higher heat transfer. In the limit case of a parallel disk, $\beta = 0$, the correlation (5.6) for a very thin stationary disk would predict a mean Nusselt number value of $Nu = 535$. The present experimental data agree better with correlation (5.7) or (5.8) if a slightly higher value is used for the correlation constant that which is justified by the finite thickness of the employed disk, see also the discussion in Chap. 5.

At high rotation rates, flow separation can also be observed at a significantly higher incidence $\beta > \beta_{\text{tr, }\omega}$. Here, the transition value $\beta_{\text{tr},\omega}$ depends on the Reynolds number ratio Re_ω/Re_u. The corresponding flow field is schematically shown in

Fig. 6.15c. Due to the high rotation, the stagnation point moves into the flow domain for $\beta > \beta_{\mathrm{tr},\omega}$, and a comparably high mean heat transfer results due to the flow separation. Its value is of the same magnitude as in turbulent flow past a parallel blunt disk.

A first noticeable deviation (increase) from the stagnation flow heat transfer occurs at an onset value $\beta_{\mathrm{tr},\omega+}$. The experimentally observed value $\beta_{\mathrm{tr},\omega+}$ can be correlated by an empirical expression

$$\beta_{\mathrm{tr},\,\omega+} = \beta_{\mathrm{tr}} + \exp(1.545 Re_\omega / Re_u), \tag{6.22}$$

first proposed in [19]. It should be noted that the purely empirical correlation is only proven for $d/R = 0.3$ corresponding to $\beta_{\mathrm{tr}} = 26.6°$. For a Reynolds number ratio $Re_\omega / Re_u = 2$, expression (6.22) gives an onset value $\beta_{\mathrm{tr},\omega+} = 48.6°$. In the case of vanishing rotation rates, expression (6.22) reduces directly to the subcritical bifurcation value β_{tr}. The transition value $\beta_{\mathrm{tr},\omega} < \beta_{\mathrm{tr},\omega+}$ is defined by the sharp step in the mean Nusselt number, see Fig. 6.14. For $Re_\omega / Re_u = 2$, a value $\beta_{\mathrm{tr},\omega} = 45° \pm 0.5°$ was experimentally observed. The transition value $\beta_{\mathrm{tr},\omega}$ can be predicted by a simplified model resting on the stagnation point location as a function of inflow velocity u_∞ and rotation ω of an inclined blunt disk with radius R. This model is an extension of the potential flow discussion of Sect. 6.3. Due to the angle of attack β, the effective circulation is $\Gamma \cos \beta$ instead of Γ, as in the case of a parallel disk or ellipsoid. The matching condition leads then to

$$\frac{\omega R \cos \beta}{u_\infty} = \frac{u_M}{u_\infty} = R_{\mathrm{cr}} \Leftrightarrow \frac{Re_\omega}{Re_u} = \frac{R_{\mathrm{cr}}}{\cos \beta_{\mathrm{tr},\,\omega}}. \tag{6.23}$$

Assuming the critical value $R_{\mathrm{cr}} = 1.4$ obtained for the parallel disk (see the discussion of the Landau model in Sect. 6.3), expression (6.23) predicts a transition value of $\beta_{\mathrm{tr},\omega} = 45.6°$ for $Re_\omega / Re_u = 2$. This value agrees extremely well with the experimentally observed value. Obviously, a subcritical transition at $\beta = \beta_{\mathrm{tr},\omega}$ should only exist for $Re_\omega / Re_u > R_{\mathrm{cr}}$. In the case of $Re_\omega / Re_u = R_{\mathrm{cr}}$, the above transition is predicted to be at $\beta_{\mathrm{tr},\omega} = 0°$ (i.e., parallel disk).

A close inspection of the experimental data shown in Fig. 6.14 indicates that the simplified model above is far from being absolutely accurate, but it helps in understanding how the flow parameters qualitatively govern the mean heat transfer.

References

1. Mabuchi I, Tanaka T, Sakakibara Y (1967) Studies on the convective heat transfer from rotating disk (1st report, the effect of dissipative energy on the laminar heat transfer from a disk rotating in uniform forced stream). Bull JSME 10:104–112
2. Mabuchi I, Tanaka T, Kumada M (1968) Studies on the convective heat transfer from rotating disk (4th report, laminar heat transfer from a disk with a step-wise discontinuity in surface temperature rotating in uniform forced stream). Bull JSME 11:885–893

3. Mabuchi I, Tanaka T, Sakakibara Y (1971) Studies on the convective heat transfer from rotating disk (5th report, experiment on the laminar heat transfer from a rotating disk in a uniform forced stream). Bull JSME 14:581–589
4. Helcig C, aus der Wiesche S (2013) The effect of the incidence angle on the flow over a rotating disk subjected to forced air streams. In: Proceedings ASME fluids engineering summer meeting, Incline Village, Nevada (paper FEDSM2013-16360)
5. Trinkl CM, Bardas U, Weyck A, aus der Wiesche S (2011) Experimental study of the convective heat transfer from a rotating disc subjected to forced air streams. Int J Therm Sci 50:73–80
6. Cobb EC, Saunders OA (1956) Heat transfer from a rotating disk. Proc Roy Soc A 236:343–351
7. Dennis RW, Newstead C, Ede AJ (1970) The heat transfer from a rotating disc in an air crossflow. In: Proceedings of 4th international heat transfer conference, Paris-Versailles (paper FC 7.1)
8. aus der Wiesche S (2002) Heat transfer and thermal behaviour of a rotating disk passed by a planar air stream. Forsch Ingenieurwes 67:161–174
9. aus der Wiesche S (2004) LES study of heat transfer augmentation and wake instabilities of a rotating disk in a planar stream of air. Heat Mass Transf 40:271–284
10. aus der Wiesche S (2007) Heat transfer from a rotating disk in a parallel air crossflow. Int J Therm Sci 46:745–754
11. Tobak M, Peake DJ (1982) Topology of three-dimensional separated flows. Annu Rev Fluid Mech 14:61–85
12. Landau LD, Lifshitz EM (1990) Hydrodynamik. Akademie, Berlin
13. Booth GL, de Vere APC (1974) Convective heat transfer from a rotating disc in a transverse air stream. In: Proceedings of 5th international heat transfer conference, Tokyo (paper FC1.7)
14. Latour B, Bouvier P, Harmand S (2011) Convective heat transfer on a rotating disk with transverse air crossflow. ASME J Heat Transf 133, paper-ID 021702 (10 p)
15. Milne-Thomson LM (1996) Theoretical hydrodynamics. Dover, New York
16. Smirnow WI (1987) Lehrgang der Höheren Mathematik. Band III/2. VEB Verlag der Wissenschaften, Berlin
17. Rott N, Lewellen WS (1967) Boundary layers due to the combined effects of rotation and translation. Phys Fluids 10:1867–1873
18. aus der Wiesche S (2010) Heat transfer from rotating discs with finite thickness subjected to outer air streams. In: Proceedings of 14th international heat transfer conference, Washington, DC, (paper IHTC14-22558)
19. Helcig C, aus der Wiesche S (2014) Convective heat transfer from a free rotating disk subjected to a forced crossflow. In: Proceedings ASME Turbo Expo, Düsseldorf, Germany (paper GT2014-26223)
20. Helcig C, aus der Wiesche S, Shevchuk IV (2014) Internal symmetries, fundamental invariants, and convective heat transfer from a rotating disk. In: Proceedings 15th international heat transfer conference (Begell House Digital Library), Kyoto (paper IHTC15-22558)

Chapter 7
Large-Eddy-Simulation (LES) Analysis

Computational Fluid Dynamics, usually abbreviated as CFD, has become a "third" approach in addition to the classic analytical treatment and the experimental investigation of flow and heat transfer phenomena. CFD is a branch of fluid mechanics that uses numerical methods and mathematical algorithms to solve and analyze problems that involve flow phenomena. This approach is especially attractive since powerful computers for performing the calculations are now widely available. However, the direct numerical simulation (DNS) of turbulent flows resolving the entire range of turbulent length scales at high Reynolds numbers is still not feasible, and appropriate simulation strategies for such flows are still required.

In industry and in many research institutions, the Reynolds averaged Navier–Stokes (RANS) equation approach dominates. It is the oldest approach to turbulence modeling. An ensemble version of the governing equations is solved, which introduces new apparent stresses known as Reynolds stresses. This approach adds a second-order tensor of unknowns for which various models can provide different levels of closure. Details about the approach can be found in a large number of books, e.g. [1–3]. RANS simulation can be extremely powerful for calculating time-averaged flow quantities in statistically stationary turbulent flows. On the other hand, many real flows are characterized by major unstationary phenomena such as vortex shedding. In these cases, the classic RANS approach is not sufficient, and substantial efforts are required to extend the RANS formalism. In 1971 Reynolds and Hussain [4] introduced such formalism based on phase-averaging of the instantaneous velocity field with respect to the period of the shedding. Their approach may justify unstationary numerical solutions, but phase-averaging is not well-defined for flows where coherent vortices are unpredictable, as in many shear flows.

Currently, wide use is made of the so-called unstationary RANS methods (or URANS) for complex turbulent flows [4]. URANS models are also offered by many commercial CFD codes. The fundamental problem is that URANS solutions

© The Author(s) 2016
S. aus der Wiesche, C. Helcig, *Convective Heat Transfer From Rotating Disks Subjected To Streams Of Air*, SpringerBriefs in Applied Sciences and Technology, DOI 10.1007/978-3-319-20167-2_7

are frequently not rigorously justified, and the significance of the so-computed solutions cannot be checked. It will be shown in this chapter that URANS simulations can be interpreted as loosely resolved large-eddy-simulations (LES), as remarked by Lesieur et al. [5]. LES is a technique in which the smallest scales of the flow are removed through a filtering operation, and their effect is modeled using subgrid-scale models. The history of LES began in 1963 with the proposal of the later-called Smagorinsky eddy viscosity model [6]. Smagorinsky wanted to represent the effects on quasi-two-dimensional atmospheric large-scale flow of three-dimensional subgrid-scale turbulence. This approach was extremely fruitful for a lot of applications after the advent of powerful computers in the 1990s. This technique allows the largest and most important scales of the turbulence to be resolved, while reducing the computational cost. The method requires substantially greater computational resources than RANS methods but is far cheaper than DNS. It is not possible to cover all LES aspects in a single chapter; LES has become a discipline to itself [5]. LES is considered here as a powerful tool for investigating in greater detail flow and heat transfer phenomena involved in rotating disk systems. Since important LES details are still far from being commonly understandable for many users, the present chapter contains basic concepts of LES for incompressible Newtonian flows with attention being paid to its application to rotating disk flows. Based on the brief presentation, important representative examples of LES analyses are discussed. LES is particularly useful for investigating the spatiotemporal variations in turbulent boundary layer flows, for which the simple empirical correlations or RANS methods do not offer an adequate approach.

7.1   LES Overview and Formalism

Understanding LES is not possible without considering the Kolmogorov spectrum theory [7]. In turbulent flows, different length scales can be distinguished. The largest scale is comparable with the major dimensions of the considered geometry, and the smallest scale is the Kolmogorov dissipative scale. There is also some evidence that, in comparison to large-scale turbulence, small-scale turbulence is not far from isotropy. This fact means that considerations justified for the special case of isotropic turbulence can also be exploited for complex flows. The turbulent energy density $E(k)$ decreases with an increasing spatial wave number k because it results from a process in which the dynamic structures of size $1/k$ (called eddies) interact with each other and with the more energetic large-scale eddies to generate smaller eddies. These smaller eddies populate eddies of yet smaller sizes, and energy is transferred from the large to the small scales. This process is called inertial energy cascade or turbulent cascade and it works without dissipation. This cascade terminates at small scales where an eddy is so small that it diffuses appreciably as a result of viscosity.

A DNS could provide an exact solution if the numerical schemes were accurate enough and all scales of motion were captured. At present, even the most powerful supercomputers are not able to resolve the smallest scales for flows at higher Reynolds numbers, which occur in technical applications, because computing times and memory requirements are too excessive. LES models are attractive because they do not try to resolve all scales. In fact, LES resolves only the largest scales. Rigorously spoken, the popular term "large-eddy-simulation" is a little bit unlucky; the German or French expressions meaning "large-scale simulations" ("Grobstruktursimulationen" or "simulation des grandes échelles", respectively) are slightly more appropriate.

The principal operation in large-eddy-simulation LES is *low-pass filtering*. This operation is applied to the Navier–Stokes equations to eliminate small scales of the solution. This approach reduces the computational cost of the simulation in comparison to DNS. The governing equations are thus transformed, and the solution is a *filtered* velocity field. The larger scales are captured by LES, and hence LES simulations are always unsteady and three-dimensional. This is somewhat in contrast to RANS equations which are obtained by *ensemble-averaging* from the Navier–Stokes equations. RANS simulations target the time-averaged flow quantities and therefore RANS simulations can also be performed employing two-dimensional schemes. LES analyses provide both instantaneous and statistical data. Therefore postprocessing of LES analyses requires substantially more effort than that of RANS simulations.

In this book, the fluid (air) is treated as an incompressible Newtonian fluid with practically constant material properties, see also Chap. 2. The governing flow field equations are hence given by the Navier–Stokes equations (2.1) and the continuity equation (2.2). In addition, the energy equation also has to be considered in heat transfer problems. Initially, LES formalism was developed for the physical space. This approach is still the most common in practice, and is presented in this chapter, but it should be kept in mind that a spectral LES formalism is also available for isotropic turbulence [5].

The scale characteristic of the employed computational grid mesh is denoted by Δx. To eliminate the unresolved subgrid-scale effects, a filter of width Δx and a filter function $G_{\Delta x}$ are employed. Mathematically, the filtering operation corresponds to the convolution of any flow field quantity $f(\mathbf{x}, t)$ by the filter function $G_{\Delta x}(\mathbf{x})$ in the form

$$\overline{f}(\mathbf{x}, t) = \iiint f(\mathbf{y}, t) G_{\Delta x}(\mathbf{x} - \mathbf{y}) d\mathbf{y}. \tag{7.1}$$

The instantaneous (exact) flow field quantity f can thus be expressed by a filtered contribution and a subgrid-scale contribution f', i.e.,

$$f = \overline{f} + f'. \tag{7.2}$$

Assuming Δx as constant, the filtering operator commutes with space and time derivatives of the governing equations. The filtered continuity is therefore

$$\frac{\partial \bar{u}_i}{\partial x_i} = 0. \tag{7.3}$$

The filtered momentum equation is

$$\frac{\partial \bar{u}_i}{\partial t} + \frac{\partial}{\partial x_j}\left(\bar{u}_i \bar{u}_j\right) = -\frac{1}{\rho}\frac{\partial \bar{p}}{\partial x_i} + \frac{\partial}{\partial x_j}\left(2\nu \bar{S}_{ij} + T_{ij}\right) \tag{7.4}$$

in the usual tensor notation. In (7.4), $\nu = \mu/\rho$ is the molecular kinematic viscosity,

$$\bar{S}_{ij} = \frac{1}{2}\left(\frac{\partial \bar{u}_i}{\partial x_j} + \frac{\partial \bar{u}_j}{\partial x_i}\right) \tag{7.5}$$

and

$$T_{ij} = \bar{u}_i \bar{u}_j - \overline{u_i u_j} \tag{7.6}$$

is the subgrid-stresses tensor responsible for momentum exchanges between the subgrid and the filtered scales. For a passive variable θ (e.g., temperature), the corresponding filtered equation is

$$\frac{\partial \bar{\theta}}{\partial t} + \frac{\partial}{\partial x_j}\left(\bar{\theta} \bar{u}_j\right) = \frac{\partial}{\partial x_j}\left(\alpha \frac{\partial \bar{\theta}}{\partial x_j} + q_j\right) \tag{7.7}$$

with the subgrid-flux

$$q_j = \bar{\theta} \bar{u}_j - \overline{\theta u_j}. \tag{7.8}$$

On the face of it, the set of equations (7.3) and (7.4) look very similar to the frequently employed URANS equations, but their physical meanings are quite different. In LES, the flow quantities are calculated after filtering whereas averaged quantities are considered for RANS.

In LES, the subgrid-stresses tensor T_{ij} and flux q_j have to be modeled in a manner that is consistent with the turbulence theory. For that, it is convenient to express the subgrid terms in accordance with

$$T_{ij} = -\left(\overline{u_i' u_j'} + \overline{\bar{u}_i u_j'} + \overline{\bar{u}_j u_i'} + \overline{\bar{u}_i \bar{u}_j} - \bar{u}_i \bar{u}_j\right) \tag{7.9}$$

$$q_j = -\left(\overline{\theta' u_j'} + \overline{\bar{\theta} u_j'} + \overline{\bar{u}_j \theta'} + \overline{\bar{\theta} \bar{u}_j} - \bar{\theta} \bar{u}_j\right). \tag{7.10}$$

In (7.9), the first term of the right-hand side is the Reynolds-stress-like term, and the last two terms are the Leonard tensor [8]. The latter is defined in terms of the filtered field variables, and it has been used in scale-similarity models to provide information on the subgrid stresses. The terms containing the subgrid contributions need to be modeled as for RANS.

7.2   Eddy Viscosity and Diffusivity Assumption

By analogy with RANS, the subgrid-scale tensors are generally expressed in terms of eddy viscosity and diffusivity coefficients ν_t and α_t in the mathematical form

$$T_{ij} = 2\nu_t(\mathbf{x}, t)\overline{S}_{ij} + \frac{1}{3}T_{ll}\delta_{ij}, \tag{7.11}$$

$$q_j = \alpha_t(\mathbf{x}, t)\frac{\partial\overline{\theta}}{\partial x_j}. \tag{7.12}$$

Then, the LES equations for an incompressible Newtonian flow can be written as

$$\frac{\partial\overline{u}_i}{\partial t} + \frac{\partial}{\partial x_j}\left(\overline{u}_i\overline{u}_j\right) = -\frac{1}{\rho}\frac{\partial\overline{p}^*}{\partial x_i} + \frac{\partial}{\partial x_j}\left(2(\nu + \nu_t)\overline{S}_{ij}\right) \tag{7.13}$$

and

$$\frac{\partial\overline{\theta}}{\partial t} + \frac{\partial}{\partial x_j}\left(\overline{\theta}\overline{u}_j\right) = \frac{\partial}{\partial x_j}\left((\alpha + \alpha_t)\frac{\partial\overline{\theta}}{\partial x_j}\right) \tag{7.14}$$

in addition to the continuity equation (7.3). Due to the model (7.11), the trace T_{ll} of the subgrid stresses tensor is unknown. Therefore, only the macro-pressure

$$\overline{p}^* = \overline{p} - \frac{1}{3}\rho T_{ll} \tag{7.15}$$

can be calculated by means of the above set of equations. If there is no interest in the absolute value of the static pressure, the lack in modeling is not substantial.

The eddy-viscosity assumption poses some fundamental questions. First of all, low-pass filtering of DNS results indicated that the classic eddy-viscosity concept in physical space might not be rigorously justified [5]. In this case, determining the turbulent Prandtl number

$$Pr_t = \frac{\nu_t}{\alpha_t} \tag{7.16}$$

can be questioned. Frequently, the simplified assumption $Pr_t = 1$ is done (see later examples), and then there is only a need to model the eddy-viscosity ν_t.

The most widely used eddy-viscosity model is the Smagorinsky model proposed in 1963 in the pioneering LES paper [7]. It is practically an adaption of Prandtl's mixing length theory to subgrid-scale modeling. Prandtl assumed that the eddy viscosity (arising in RANS equations) is proportional to a turbulent scale (the mixing length) multiplied by a turbulent characteristic velocity [3]. Smagorinsky assumed the subgrid-scale Δx as a characteristic length scale, and the characteristic velocity is given by means of $\Delta x|\overline{S}| = \Delta x\sqrt{2\overline{S}_{ij}\overline{S}_{ij}}$. Thus, the eddy-viscosity is

$$\nu_t = (C_S \Delta x)^2 |\overline{S}|. \tag{7.17}$$

The Smagorinsky constant C_S is the subject of debates. As demonstrated by Lilly in 1967 [9], its value can be calculated in isotropic turbulence by means of an argument based on Kolmogorov's cascade. Considering a value of 1.4 for the Kolmogorov constant, the value of the Smagorinsky constant is $C_S = 0.18$. This value is appropriate for LES of isotropic turbulence, but it has been proven to yield inaccurate results for other flows. It yielded a too dissipative model in the presence of walls. Many researchers prefer an empirical value of $C_S = 0.1$, for which Smagorisnky's model provides satisfying LES results for free-shear flows and for channel flows. An assessment of corresponding questions regarding subgrid-modeling has been carried out by Rodi et al. [10]. It should be noted that so far no "standard" LES model has been proposed. Instead, it can be shown that each LES case is a priori unique: both the topology of the computational grid and the details of the numerical method enter the simulation error definitions [11].

Another serious issue with the classic Smagorinsky model with regard to many applications is its artificial relaminarization of the flow if the upstream perturbation is not high enough. This problem is discussed in detail by Meneveau and Katz [12]. Assuming close to the wall $u_1 \propto y^+$, $u_2 \propto y^{+2}$, Smagorinsky's model yields a finite value for the component T_{12} of the subgrid-scale tensor T_{ij}. This observation justifies the development of dynamic models initiated by Germano [13]. The underlying idea of dynamical models is to extract information concerning a given eddy-viscosity model via a *double-filtering* in physical space. Details about that approach can be found elsewhere [5, 13]. It can be shown that the dynamic model gives a zero subgrid-scale stress at a fixed wall, and it leads to a better asymptotic behavior near the wall. On the other hand, the dynamic model can introduce a destabilizing mechanism in numerical simulations because it allows a sort of backscatter in physical space (i.e., negative values of the eddy-viscosity are possible).

Finally, the concept of so-called detached eddy simulation (DES) should be mentioned in this section, too. DES can be interpreted as a modification of RANS modeling in which the model switches to a subgrid-scale formulation in regions fine enough for LES calculations. Regions near solid walls and where the turbulent length scale is less than the maximum grid dimension are treated by the RANS solution method. Where the turbulent length scale exceeds the grid dimension, the regions are solved by means of the LES method. The advantage of this concept is that the grid resolution is not as demanding as in the case of pure LES, but grid generation is more complicated than for a simple RANS or LES due to the inherent switch between RANS and LES. DES provides a single smooth velocity field across both the RANS and the LES regions of the solution. Further details about this topic can be found elsewhere [14, 15]; in the present chapter, RANS or DES models will not be further explicitly discussed.

7.3 Application of LES for Basic Flows for Rotating Disks

Though the flow configuration of free or enclosed rotating disk systems is often axisymmetric, experiments have revealed the existence of large-scale vertical structures in the turbulent regime [16]. Even the basic flow past a free rotating disk exhibits interesting features that cannot be resolved by the classic RANS approach. Inside the laminar boundary layer near the rotating disk, all velocity components are non-zero. As discussed earlier, the flow over a rotating disk is laminar for local Reynolds numbers $Re_{\omega,r} = \omega\, r^2/v$ of less than about 4.5×10^4 [17]. The flow is fully turbulent for $Re_{\omega,r}$ greater 4×10^5. Lingwood' stability analysis [18] shows how the onset of transition over a rotating disk occurs at $Re_{\omega,\delta2}$ above 502 and below 513 where δ_2 is the boundary layer momentum thickness chosen as the characteristic length scale for the definition. Typically, three-dimensional turbulent boundary layers are formed from initially two-dimensional flow due to action of spanwise pressure gradients or shearing forces. The turbulent boundary layer flow past a rotating disk represents a unique exception because it directly starts three dimensionally. Its underlying structure does not result from a perturbation of an initially two-dimensional flow. Therefore, the flow past a rotating disk offers an ideal starting point for investigations into the underlying structures of three-dimensional turbulent boundary layers, and valuable experimental data have been obtained in this area by Eaton and coworkers [19, 20].

In 2000, Wu and Squires performed an LES analysis of the turbulent flow over a rotating disk [21]. The primary aims of their work were the numerical prediction of the three-dimensional turbulent boundary layer and an investigation of the underlying flow structure. The considered Reynolds number was $Re_{\omega,r} = 6.5 \times 10^5$ (corresponding to $Re_{\omega,\delta2} = 2660$). Since it was difficult to incorporate the laminar-to-turbulent transition into the calculation, only a suitable segment (radial and azimuthal) was chosen as the computational domain for the fully turbulent flow. The use of such a reduced domain requires special attention with regard to the boundary conditions [21]. Wu and Squires formulated the governing equations in cylindrical coordinates, and they employed three subgrid models, namely the dynamic eddy viscosity model of Germano et al. [22], the dynamic mixed model of Zang et al. [23], and the dynamic mixed model of Vrenan et al. [24]. The classic Smagorinsky model was not considered because prior experiences demonstrated a better performance of the dynamic models.

Wu and Squires realized that the prediction of complex flows using LES requires special care with regard to the numerical scheme. Upwind methods introduce a dissipative truncation error that can act as an additional (unphysical) subgrid model. The governing equations including the subgrid models were solved using a semi-implicit fractional step method in cylindrical coordinates. Second-order central differences were used for approximating spatial derivatives on a staggered grid. The numerical scheme was essentially the same as that used by Akselvoll and Moin [25]. A series of calculations were performed to validate the overall computational approach. Such an effort is typical for LES analyses because the somewhat empirical input of the subgrid models and the numerical scheme must be carefully considered before the results from an LES calculation may be used. It was found that when the resolution was in a range in which large-scale motions were accurately and well resolved, the effect of the subgrid models and grid refinement were not significant. The LES predictions show that the skewing angle of the wall shear stress (relative to the disk velocity) was about 16°.

Wu and Squire provided a detailed comparison of LES predictions with experimental measurements [19] of the turbulence quantities, and generally a good agreement was found. The numerical results supported the structural model advanced by Littell and Eaton [19], where streamwise vortices with the same sign as the streamwise vorticity are mostly responsible for strong sweep events whereas streamwise vortices having the opposite sign promote strong ejections. Near-wall characteristics of the turbulence intensities and turbulent shear stress in the disk flow were presented and compared with available DNS results [26, 27]. They concluded that these disk flow results are not markedly different from their counterparts in canonical two-dimensional flows.

Enclosed rotating disk systems and rotor-stator configurations were numerically investigated by several authors employing LES. Lygren and Anderson [28, 29] conducted an LES study that considered an axisymmetric and statistically steady-state turbulent flow within an angular section of an unshrouded rotor-stator cavity for a rotational Reynolds number of order $Re_\omega = 1.6 \times 10^6$. Their results indicated that the mixed dynamic subgrid-scale model of Vreman et al. [24] provided better results compared to Lilly's dynamic subgrid-scale model [30].

In addition to the LES formulations in physical space, spectral approaches have been considered with success in cases of rotating disk systems, too. For an enclosed rotor-stator system, the spectral vanishing viscosity technique by Severac and Serre [31] provided excellent results.

7.4 Rotating Disk Heat Transfer LES

The above brief discussion of LES applications to flows involving rotating disk indicates the potential of this modeling approach for covering turbulent phenomena. For heated rotating disk systems, both the temperature field and the flow field have to be resolved numerically. For LES modeling, this task directly poses the question as to the value of the turbulent Prandtl number (7.16).

A LES study [32] of heat transfer augmentation and wake instabilities of a rotating disk in a planar stream of air was presented in 2004. The schematic of the considered flow problem is sketched in Fig. 7.1. In study [32], a simple box filter and the classic Smagorinsky model were used to model the eddy viscosity. A value of $C_S = 0.1$ was chosen for the Smagorinsky constant, and a constant turbulent Prandtl number $Pr_t = 1$ was assumed. Upwind differencing for the convective terms was employed. A very thin disk was used with constant surface temperature with radius $R = 0.1$ m and thickness $d = 0.002$ m. It was placed in the center of the

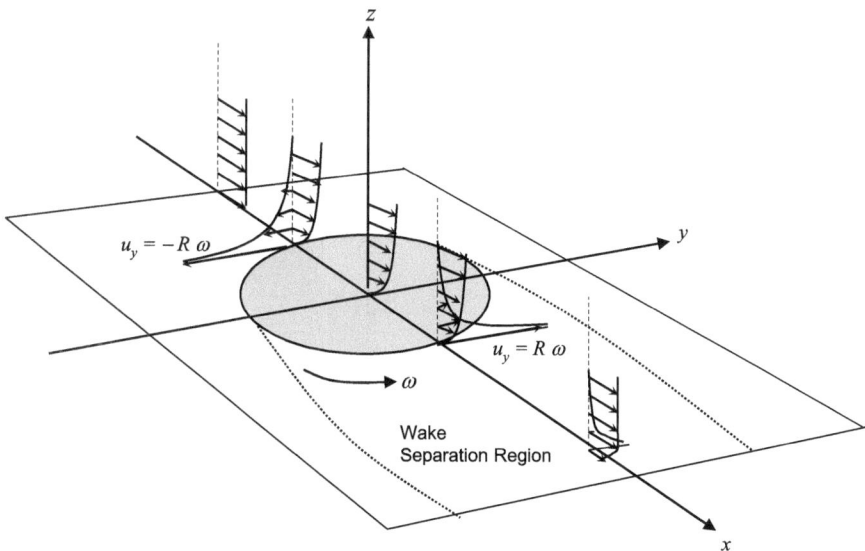

Fig. 7.1 Schematics of the flow problem considered in the LES study [32]

computational domain and a uniform stream of air at atmospheric conditions was prescribed at the inlet corresponding to a translational Reynolds number $Re_u = u_\infty R/\nu = 6666$. Two values of disk running speed ω were considered corresponding to $Re_\omega = 2\,Re_u$ and $10\,Re_u$. The main objective of the case study [32] was to gain insight into the vortex dynamics at the vicinity of the bifurcation point discussed earlier. As discussed in Chap. 6, a mean heat transfer augmentation occurred for a critical Reynolds number ratio Re_ω/Re_u. The mean Nusselt number Nu_m as a function of the involved Reynolds number ratio can be well described by a phenomenological Landau model, but such a model is not able to explain the underlying flow mechanism.

The above LES modeling assumptions and the numerics may be criticized in that the subgrid-scale model they employ is too dissipative. Furthermore, the considered disk dimensions do not permit a comparison with available experimental results. On the other hand, the LES study [32] related the observed flow instability and bifurcation (Landau model) to a periodic vortex generation process and indicated an abrupt transition to turbulence. Some illustrations of the computed instantaneous thermal wakes are shown in Fig. 7.2, where the qualitative different vortex structures in the wakes become obvious.

The number of considered flow cases (i.e., Reynolds number ranges) was substantially increased in a subsequent LES study [33] performed by the same author. In that study, the disk radius was also increased ($R = 0.2$ m), but no substantial changes were made with regard to the subgrid-scale modeling and the numerics. Furthermore, the considered isothermal disk remained very thin. Therefore the results of [32, 33] have more academic than applied value, as pointed out by Shevchuk [34]. In respect to the convective heat transfer from a free rotating disk

Fig. 7.2 Computed thermal
wake of a thin rotating
isothermal disk subjected to
a uniform stream of air for
two different rotational
Reynolds numbers [32]

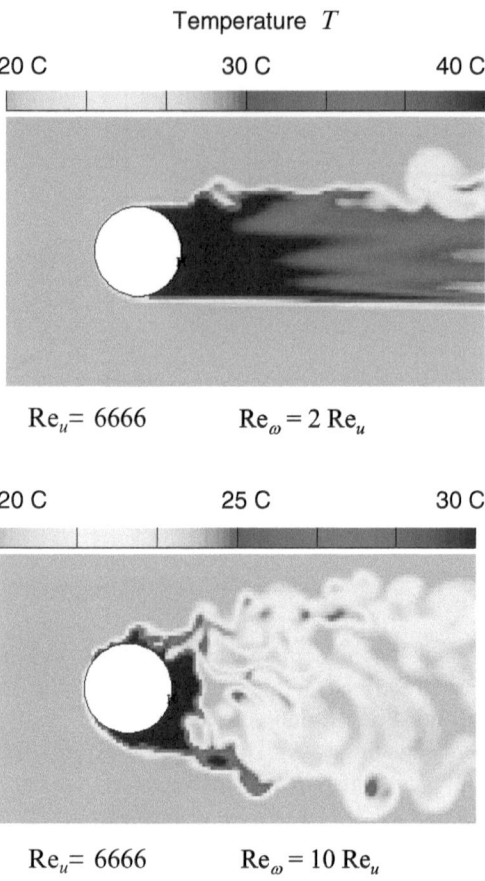

Temperature T

20 C   30 C   40 C

$$Re_u = 6666 \qquad Re_\omega = 2\,Re_u$$

20 C   25 C   30 C

$$Re_u = 6666 \qquad Re_\omega = 10\,Re_u$$

without any crossflow, the LES predictions [33] for the mean Nusselt number Nu_m
were in excellent agreement with the theoretical results. An asymptotic correlation

$$Nu_\mathrm{m} = 0.015\,Re_\omega^{0.8}. \qquad (7.18)$$

was predicted for the fully turbulent regime, which also agrees well with the literature
[34]. On the other hand, the prediction of the convective heat transfer from a disk
subjected to a parallel stream of air was too low in the LES studies [33, 35]. One
reason for this failure can be identified as the underpredicted flow separation
and reattachment of a turbulent boundary with a blunt disk, and another issue can be
attributed to the comparably poor performance of the employed simple Smagorinsky
model, particularly at the disk surface. However, the bifurcation phenomenon and the
corresponding vortex dynamics in the wake were sufficiently resolved by the LES
study [33]. Furthermore, the phenomenon that a constant value of an alignment angle
of order $\theta_s = 12.5 \pm 0.5°$ exists (see Chap. 6, Sect. 6.4) were well reproduced by
the predicted local Nusselt number distribution on the disk surface.

A certainly more appropriate subgrid-scale modeling was chosen by Tuliszka-Sznitko and coworkers in LES studies that considered rotor-stator disk systems [36, 37]. The objective of those studies was to analyze the coherent structures of transitional and turbulent flows and to compute statistical flow and heat transfer parameters. The numerical solution was performed by means of a pseudospectral collocation Chebyshev–Fourier–Galerkin approximation. The LES employed a version of the dynamic Smagorinsky eddy viscosity model proposed by Meneveau et al. [38]. In this approach, the stabilizing averaging is accumulated over the fluid pathlines instead of averaging over the direction of statistical homogeneity. The Smagorinsky coefficient is dynamically determined by minimizing the modeling error over the fluid path lines. The turbulent Prandtl number Pr_t was calculated and compared with available experimental data. It was found that its value is between 0.8 up to 1.2, and hence the frequently used assumption of $Pr_t = 1$ for gases might be somewhat justified.

The advantage of using a dynamic model is demonstrated by means of Figs. 7.3 up to 7.6 where numerical results are shown obtained for a thick rotating disk subjected to a parallel stream of air. The LES analysis was performed using a finite-volume method for an incompressible fluid with constant material properties. In addition to the flow quantities, velocity and pressure, a passive scalar variable (i.e., temperature) was solved, too. The discretization in space and time was of second order using a central differencing scheme. The subgrid stress tensor and the subgrid contribution to the scalar were modeled in terms of filtered quantities using the Lagrangian dynamic mixed model (LDMM) adapted from the original Lagrangian

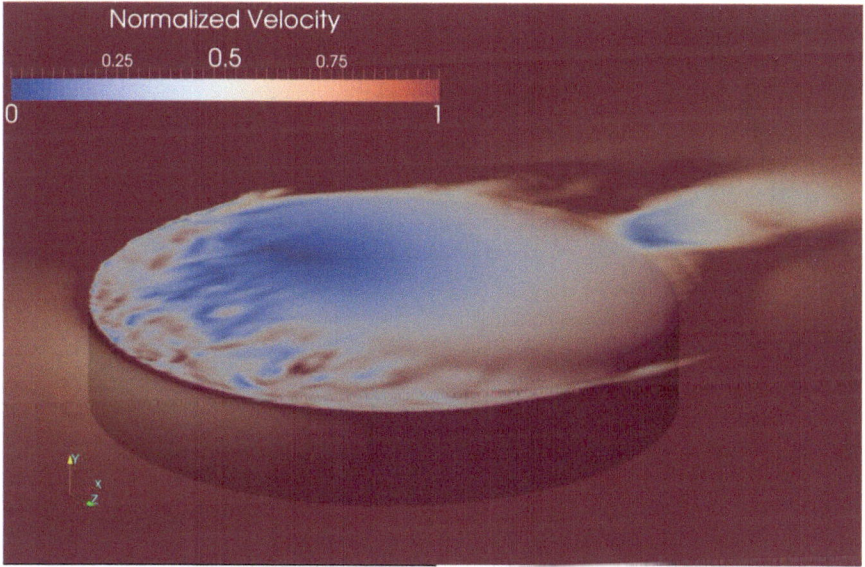

Fig. 7.3 Calculated instantaneous velocity field over a parallel rotating disk at $Re_\omega = 4000$ and $Re_u = 4000$

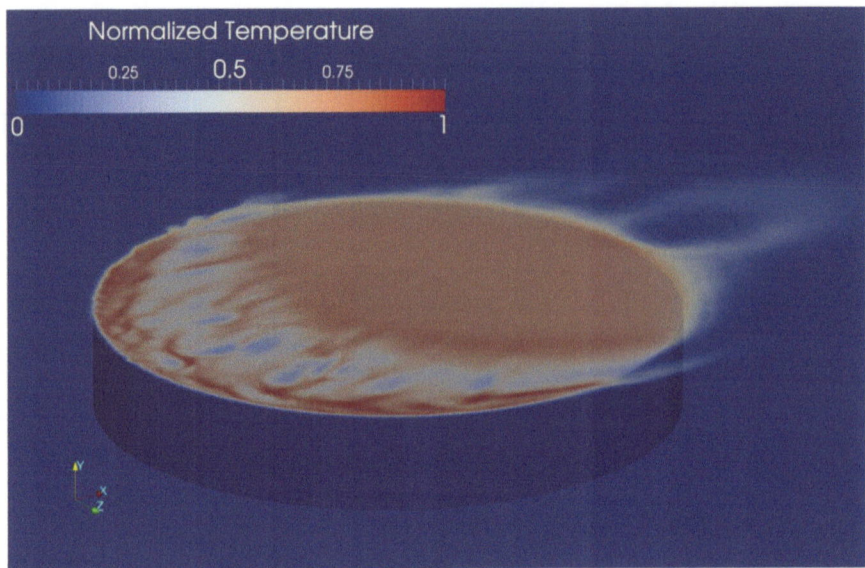

Fig. 7.4 Calculated instantaneous normalized temperature distribution over a parallel rotating disk at $Re_\omega = 4000$ and $Re_u = 4000$

dynamic model proposed by Meneveau et al. [38]. The turbulent viscosity and the turbulent Prandtl number were determined dynamically in space and time. In contrast to the simple Smagorinsky model employed earlier in [32], the present dynamical model is able to resolve small-scale effects close to the rotating disk with a higher accuracy. The calculated normalized instantaneous velocity and temperature fields for small Reynolds numbers are shown in Figs. 7.3 and 7.4, respectively. The flow separation at the leading edge and the destruction of the symmetry with respect to the velocity distribution close to the surface due to rotation can clearly be identified in Fig. 7.3. The temperature distribution remained nearly symmetrical, Fig. 7.4, but regular substructures were calculated within the separation bubble zone. These structures might be related to Taylor-Goertler vortices. In case of substantial larger Reynolds numbers, irregular small-scale behavior of the turbulent flow becomes more important as indicated by the calculated normalized instantaneous velocity and temperature fields shown in Figs. 7.5 and 7.6, respectively.

Nguyen and Harmand reported an LES analysis [39] for the convective heat transfer from a rotating cylinder with a spanwise disk subjected to an air crossflow. This configuration has strong similarities to the canonical case of a rotating disk in a parallel stream. They used a commercial CFD code for their LES calculations, but no further details about the subgrid-scale model were stated in [39]. The translational Reynolds number was 23,560 and two rotational Reynolds numbers (54,900 and 109,800) were considered. The LES calculations were initiated instead by solutions obtained from a prior RANS simulation. The LES results were compared with experimental data (PIV). Their main purpose for employing LES calculations

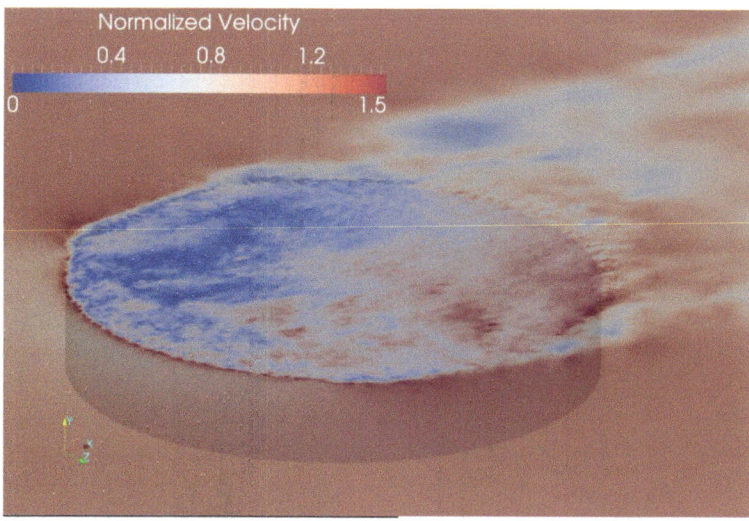

Fig. 7.5 Calculated instantaneous velocity field over a parallel rotating disk at $Re_\omega = 4.155 \times 10^5$ and $Re_u = 2.05 \times 10^5$

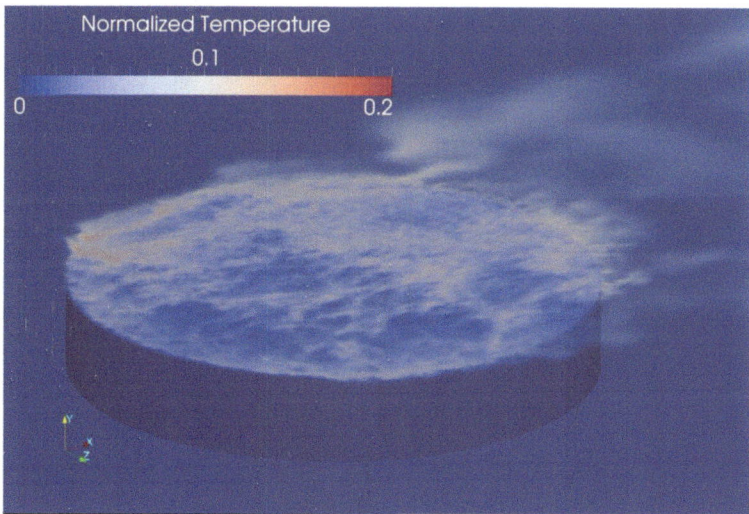

Fig. 7.6 Calculated instantaneous normalized temperature distribution over a parallel rotating disk at $Re_\omega = 4.155 \times 10^5$ and $Re_u = 2.05 \times 10^5$

was to resolve the small time-scale motions and unsteady flow structures. The time-averaged temperature and heat flux distribution on the disk surface was also shown in [39], and, as in the case of the LES analysis [33], Nguyen and Harmand also calculated a practically constant alignment angle of order $\theta_s = 12°$ to $14°$ for the half-moon-shaped surface temperature and heat flux pattern with respect to the inflow direction, but they did not comment on it in more detail. These details were not experimentally observed in a prior investigation [40]. The measurements indicated also the applicability of the Landau model proposed earlier for a free rotating disk in a parallel stream [32, 33].

7.5 Some Remarks Regarding Current Challenges for LES

The above examples demonstrate that LES offers an efficient approach for investigating turbulent flow and heat transfer phenomena for rotating disk systems in greater detail. The general issues of LES caused by its inherently high computational requirements might be decreasing due to the expected increase of future computer power, but some other fundamental issues are still challenging. Here, the numerical problems of the dynamic models resting on the eddy-viscosity approach formulated in physical space were mentioned. In principle, spectral eddy-viscosity models offer some advantages, but executing such simulations for complex geometries remains challenging. Lesieur et al. [5] provide a detailed treatment of the structure-function model, which can be interpreted as an attempt to go beyond the Smagorisnky model while keeping in physical space the same scaling as in spectral eddy-viscosity models. This approach is also available for dynamic models. With a look towards engineering applications of CFD methods, it can, however, be expected that RANS models will continue to be the "work-horse" although the use of LES methods will increase. With regard to rotating disk systems, several transient flow features cannot be accurately resolved by RANS methods. This is a good argument for increasing the use of LES methods for rotating disk systems.

References

1. Launder BE, Spalding DB (1972) Mathematical models of turbulence. Academic, London
2. Tennekes H, Lumley JL (1972) A first course in turbulence. MIT Press, Cambridge, MA
3. Rotta J (1972) Turbulente Strömungen. Teubner, Stuttgart
4. Nakayama A, Miyashita K (2001) URANS simulation of flow over smooth topography. Int J Numer Methods Heat Fluid Flow 11:723–745
5. Lesieur M, Metais O, Comte P (2005) Large-eddy simulations of turbulence. Cambridge University Press, Cambridge
6. Smagorinsky J (1963) General circulation experiments with the primitive equations. Mon Weather Rev 91:99–164
7. Lesieur M (1997) Turbulence in fluids, 3rd edn. Kluwer, Dordrecht
8. Leonard A (1974) Energy cascade in large eddy simulations of turbulent fluid flows. Adv Geophys A 18:237–248
9. Lilly DK (1967) The representation of small-scale turbulence in numerical simulation experiments. In: Proceedings IBM scientific computing symposium on environmental sciences, IBM Form 320-1951, pp 195–210

10. Rodi W, Ferziger JH, Breuer M, Pourquie M (1997) Status of large eddy simulation: results of a workshop, workshop on LES of flows past bluff bodies (Rottach-Egern, Tegernsee, Germany, 1995). ASME J Fluids Eng 119:248–262

11. Grinstein FF, Margolin LG, Rider WJ (eds) (2010) Implicit large eddy simulation. Cambridge University Press, Cambridge, Chapter 3

12. Meneveau C, Katz J (2000) Scale-invariance and turbulence models for large-eddy-simulations. Annu Rev Fluid Mech 32:1–32

13. Germano M (1992) Turbulence, the filtering approach. J Fluid Mech 238:325–336

14. Spalart PR, Jou WH, Strelets M, Allmaras SR (1997) Comments on the feasibility of LES for wings and on the hybrid RANS/LES approach. In: Advances in DNS/LES, proceedings of the first AFOSR international conference on DNS/LES

15. Strelets M (2001) Detached eddy simulation of massively separated flows. Paper AIAA 2001-0879

16. Czarny O, Iacovides H, Launder BE (2002) Precessing vortex structures in turbulent flow within rotor-stator disc cavities. Flow Turbul Combust 69:51–61

17. Kobayashi R (1994) Review: laminar-to-turbulent transition of three-dimensional boundary layers on rotating bodies. ASME J Fluids Eng 116:200–211

18. Lingwood RL (1996) An experimental study of absolute instability of the rotating disk boundary layer flow. J Fluid Mech 314:373–405

19. Littell HS, Eaton JK (1994) Turbulence characteristics of the boundary layer on a rotating disk. J Fluid Mech 266:175–207

20. Elkins CJ, Eaton JK (2000) Turbulent heat and momentum transport on a rotating disk. J Fluid Mech 402:225–253

21. Wu X, Squires KD (2000) Prediction and investigation of the turbulent flow over a rotating disk. J Fluid Mech 18:231–264

22. Germano M, Piomelli U, Moin P, Cabot WH (1991) A dynamic subgrid-scale eddy viscosity model. Phys Fluids A 3:1760–1765

23. Zang Y, Street R, Koseff JR (1993) A dynamic mixed subgrid-scale model and its application to turbulent recirculating flows. Phys Fluids 5:3186–3196

24. Vreman B, Geurts B, Kuerten H (1994) On the formulation of the dynamic mixed subgrid-scale model. Phys Fluids 6:4057–4059

25. Akselvoll K, Moin P (1996) Large-eddy simulation of turbulent confined coannular jets. J Fluid Mech 315:387–411

26. Kim J, Moin P, Moser R (1987) Turbulence statistics in fully developed channel flow at low Reynolds number. J Fluid Mech 177:133–166

27. Spalart PR (1988) Direct simulation of a turbulent boundary layer up to $Re_\theta = 1410$. J Fluid Mech 187:61–98

28. Lygren M, Andersson HI (2004) Large eddy simulations of the turbulent flow between a rotating and a stationary disk. ZAMP 55:268–281

29. Andersson HI, Lygren M (2006) LES of open rotor-stator flow. Int J Heat Fluid Flow 27:551–557

30. Lilly DK (1992) A proposed modification of the Germano subgrid-scale closure method. Phys Fluids A 4:633–635

31. Severac E, Serre E (2007) A spectral vanishing viscosity for the LES of turbulent flows within rotating cavities. J Comput Phys 226:1234–1255

32. aus der Wiesche S (2004) LES study of heat transfer augmentation and wake instabilities of a rotating disk in a planar stream of air. Heat Mass Transf 40:271–284

33. aus der Wiesche S (2007) Heat transfer from a rotating disk in a parallel air crossflow. Int J Therm Sci 46:745–754

34. Shevchuk IV (2009) Convective heat and mass transfer in rotating disk systems. Springer, Berlin

35. Trinkl CM, Bardas U, Weyck A, aus der Wiesche S (2011) Experimental study of the convective heat transfer from a rotating disc subjected to forced air streams. Int J Therm Sci 50:73–80
36. Tuliszka-Sznitko E, Zielinski A, Majchrowski W (2009) LES of the transitional flow in rotor/stator cavity. Archives Mech 61:93–118
37. Tuliszka-Sznitko E, Majchrowski W (2010) LES and DNS of the flow with heat transfer in rotating cavity. Comput Methods Sci Technol 16:105–114
38. Meneveau C, Lund TS, Cabot WH (1996) A Lagrangian dynamic subgrid-scale model of turbulence. J Fluid Mech 319:353–385
39. Nguyen TD, Harmand S (2013) Heat and mass transfer from a rotating cylinder with a spanwise disk at low-velocity crossflows. In: Proceedings ASME fluids engineering summer meeting, Incline Village, Nevada (paper FEDSM2013-16541)
40. Latour B, Bouvier P, Harmand S (2011) Convective heat transfer on a rotating disk with transverse air crossflow. ASME J Heat Transfer 133 (paper-ID 021702) (10 p)

Chapter 8
Heat Transfer Correlations for Practical Applications

The previous chapters clearly demonstrated that, in the case of rotating disks subjected to forced streams, complex flow and convective heat transfer phenomena exist. It is possible to a large extent to formulate separately adequate heat transfer correlations for the different flow and heat transfer regimes, but these correlations are limited to the considered phenomena. For instance, in the case of a stationary disk, the mean convective heat transfer can well be described by a phenomenological Landau-de Gennes model (see Chap. 5), but the presence of rotation might introduce completely new phenomena, as discussed in Chap. 6. With a look to engineering applications, it is desirable to obtain suitable heat transfer correlations that are accurate enough for practical purposes but still easy to handle for a wide class of users. This chapter discusses issues connected to that purpose.

8.1 Superposition Approach

In many convective heat transfer applications, the engineers are faced with the presence of different flow and heat transfer regimes such as laminar, turbulent, and transitional flows [1]. In many cases, simple but still accurate correlations are known but limited to a certain flow regime. An example for such a situation is given by laminar and turbulent mean heat transfer correlations for a freely rotating disk, as was discussed in Chap. 4. In detail, the mean Nusselt number Nu_m can be correlated well using a common relation

$$Nu_m = \frac{h_m R}{\lambda} = K_m \cdot Re_\omega^{n_R} \quad \text{with} \quad Re_\omega = \frac{\omega R^2}{\nu} \tag{8.1}$$

based on the rotational Reynolds number, but with different values for the constant K_m and the exponent n_R corresponding to the different flow regimes. Another

© The Author(s) 2016
S. aus der Wiesche, C. Helcig, *Convective Heat Transfer From Rotating
Disks Subjected To Streams Of Air*, SpringerBriefs in Applied Sciences
and Technology, DOI 10.1007/978-3-319-20167-2_8

similar example is given by the classic boundary layer flow past an obstacle with length scale R for which a similar set of correlations results depending on an inflow Reynolds number $Re_u = u_\infty R / v$ instead of the rotational Reynolds number. Particularly for the second example, it is often not possible to accurately predict the actual onset of transition or the actual limits of the laminar and the turbulent flow regime because the transition depends on quantities such as roughness and inflow turbulence level, which are not sufficiently known or controlled in engineering applications. With a look to such a situation, a superposition approach is widely used [2]. For a flow past a stationary obstacle, the mean Nusselt number Nu_m is simply calculated by an empirical superposition

$$Nu_m = Nu_{m,0} + \sqrt[m]{Nu_{m,l}^m + Nu_{m,t}^m} \qquad (8.2)$$

of the laminar and turbulent Nusselt numbers denoted by $Nu_{m,l}$ and $Nu_{m,t}$ Heat transfer contributions due to additional mechanisms such as natural convection or radiation can be aggregated to the empirical contribution $Nu_{m,0}$. A frequently used value for the empirical exponent m is $m = 2$. Such a correlation is recommended in the VDI Heat Transfer Atlas [2] for the convective heat transfer from stationary obstacles. The laminar and turbulent mean Nusselt numbers can typically be calculated by means of correlations given by (8.1). In principle, such a method cannot be absolutely correct because laminar and turbulent flow regimes are considered to be present simultaneously, and their effects are simply superposed. For many engineering applications, the empirical superposition approach (8.2) leads, however, to good results because the actual flow regime, and hence the corresponding heat transfer contribution, dominates in correlation due to the exponent m.

A superposition approach considering the rotational and the inflow Reynolds numbers has been proposed in [3, 4] for the mean Nusselt number Nu_m of a free rotating disk subjected to a parallel uniform stream of air. Considering the correlations for the laminar and turbulent heat transfer from a freely rotating disk and the turbulent heat transfer from a stationary disk in a uniform stream of air, the empirical correlation

$$Nu_m = \sqrt{\left(K_{m,l}Re_\omega^{0.5}\right)^2 + \left(0.036Re_u^{0.8}\right)^2} \quad \text{for} \quad Re_\omega \leq 4 \times 10^5$$

$$\qquad\qquad\qquad\qquad\qquad\qquad\qquad\qquad\qquad\qquad (8.3)$$

$$Nu_m = \sqrt{\left(K_{m,t}Re_\omega^{0.8}\right)^2 + \left(0.036Re_u^{0.8}\right)^2} \quad \text{for} \quad Re_\omega > 4 \times 10^5$$

may be used. Taking the values $K_{m,l} = 0.4$ and $K_{m,t} = 0.015$, the predictions of correlation (8.3) agree with the experimental data from Dennis et al. [5] with an error not exceeding 2.3 %. Since it is now generally accepted that the constant $K_{m,l}$ for the laminar flow over a rotating disk is better given by a value of order 0.33 up to 0.36 for air, a slight improvement of correlation (8.3) might be achieved by taking that lower value. Furthermore, it is possible to discuss the supercritical transition and heat transfer augmentation separately by using the Landau model as discussed in [4] or treated in detail in Chap. 6.

From an engineering point of view, the main overall behavior of the mean heat transfer from a parallel rotating disk is sufficiently covered by correlation (8.3), but such an approach fails in the case of inclined disks due to the bifurcation phenomena occurring at certain values for the angle of attack. Even for a stationary disk, bifurcation leads to a substantial jump of the mean Nusselt number as a function of the incidence β, as was discussed in detail in Chap. 5. Since bifurcation phenomena are fairly typical for flow and heat transfer from inclined rotating disks, the potential for the superposition approach is strongly limited there.

8.2 Recommendations for Heat Transfer Correlations

In the following, some correlations for the prediction of the mean heat transfer are recommended for practical engineering applications. It is not possible to formulate a simple correlation covering the entire range of Reynolds numbers and incidences. Instead, it is necessary to distinguish between several basic flow configurations in order to provide sufficient correlations.

Parallel rotating disk. In the case of a freely rotating disk subjected to a parallel stream of air, the superposition approach still provides a correlation that predicts the mean heat transfer with reasonable accuracy. Based on the available experimental data, the correlation

$$Nu_m = \sqrt{\left(0.33Re_\omega^{0.5}\right)^2 + \left(0.033Re_u^{0.8}\right)^2} \quad \text{for} \quad Re_\omega \leq 4 \times 10^5$$

$$Nu_m = \sqrt{\left(0.015Re_\omega^{0.8}\right)^2 + \left(0.033Re_u^{0.8}\right)^2} \quad \text{for} \quad Re_\omega > 4 \times 10^5$$

(8.4)

is recommended for the entire range of rotational Reynolds numbers and blunt disks. A performance chart of correlation (8.4) is provided in Fig. 8.1, where the predicted values are compared with available experimental data published in the literature [5–7] and own measurements. Very close to the bifurcation point where heat transfer augmentation due to rotating occurs for a supercritical Reynolds number ratio $Re_\omega/Re_u \gtrsim R_{cr} = 1.4$, the Landau model (6.6) as discussed in Chap. 6 can be used instead of the simple correlation (8.4). Since the Landau model is only applicable in the vicinity of the transition, its impact on the entire heat transfer correlation is limited, and hence the superposition approach (8.4) might be used without further modifications for parallel rotating disks subjected to a stream of air. The available experimental data are in a reasonable agreement with the predictions of correlation (8.4).

The performance chart, Fig. 8.1, indicates that correlation (8.4) has the tendency to underestimate the mean Nusselt number in comparison to experimental data.

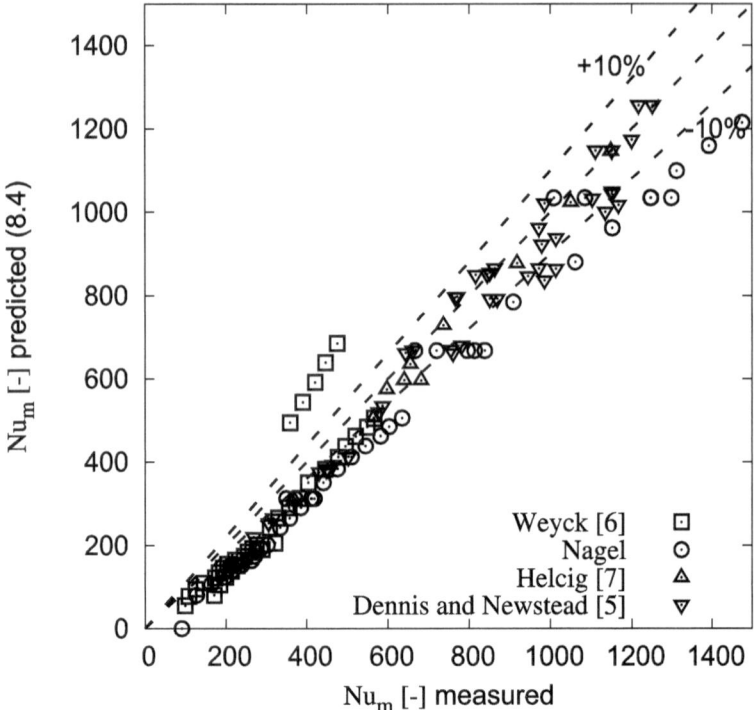

Fig. 8.1 Performance chart for heat transfer correlation (8.4) for parallel rotating disks

But it should be kept in mind that additional heat transfer contributions occurring in actual experiments lead to higher Nusselt numbers. For instance, Dennis et al. reported a constant of 0.4 instead of 0.33 for the heat transfer correlation for a rotating disk in still air [5]. It is therefore reasonable that in actual experiments, higher total heat transfer rates will be observed. Correlation (8.4) is covering only the convective heat transfer from a rotating disk subjected to a forced parallel flow. From a formal point of view, correlation (8.4) reduces to the fully turbulent correlation for a rotating disk in still air in case of sufficient high rotational Reynolds numbers. In case of moderate Reynolds numbers, the accuracy of the fully turbulent correlation is limited in case of rotating disks in still air. This is demonstrated by the series of data points (called "Weyck") plotted above the prediction. The data points correspond to actual measurements performed for a rotating disk in still air at rotational Reynolds numbers between $Re_\omega = 2 \times 10^5$ up to 4.2×10^5. In this case, the fully turbulent correlation and hence correlation (8.4) overestimates the heat transfer, because the fully turbulent regime was not reached. It is therefore recommended to use correlation (8.4) only in cases with an outer forced parallel flow.

Inclined stationary disk. In the case of an inclined stationary disk subjected to a stream of air, the superposition approach does not provide a correlation for the entire range of incidence β and inflow Reynolds number Re_u because at $\beta = \beta_{tr}$ a subcritical bifurcation occurs for laminar flow past the inclined disk. Under these conditions, the Landau-de Gennes model is recommended. In practical engineering applications, the inflow turbulence level and the inflow Reynolds number might be large enough to yield a transitional or even turbulent flow over the inclined disk. Then, there is no sharp transition at $\beta = \beta_{tr}$ that has to be considered separately.

In the case of sufficient small inflow Reynolds numbers and laminar flow over the perpendicular disk, the Landau-de Gennes model

$$Nu_m = \frac{h_m R}{\lambda} = \begin{cases} 0.63 \cdot Re_u^{1/2} & \text{for } \beta \geq \beta_{tr} \\ C_1 + C_2 \sqrt{1 - C_3(\beta - \beta_{tr})} & \text{for } \beta < \beta_{tr} \end{cases}. \tag{8.5}$$

is recommended. The value β_{tr} of the transitional incidence depends on the disk thickness ratio d/R and can be calculated by

$$\beta_{tr} = 60.6° \left(\frac{d}{R}\right)^{1/2} - 21.9° \left(\frac{d}{R}\right) \tag{8.6}$$

In principle, the constants C_1, C_2, and C_3 could be obtained on the basis of the Landau-de Gennes model and the fact that in the limit case of $\beta = 0$ the parallel stationary disk would be reached. However, this procedure would lead to rather complex functions depending on the disk thickness ratio d/R, the profile of the leading edge, and the inflow turbulence level. A performance chart of correlation (8.5) is provided in Fig. 8.2, where the predicted values for $\beta > \beta_{tr}$ are compared with available experimental data [6, 7] and own measurements. The available experimental data are in a reasonable agreement with the predictions of correlation (8.5) considering (8.6). Since only cases with $\beta > \beta_{tr}$ were considered in Fig. 8.2, correlation (8.5) is practically the Beg correlation (see Chap. 4). In case of $\beta \leq \beta_{tr}$, the resulting mean heat transfer can be roughly estimated by means of the limit case of correlation (8.4).

Inclined rotating disk. In the case of an inclined rotating disk, it is not possible to formulate a simple correlation reliable for the entire range of Reynolds numbers and inclination angles. Based on the experimental and theoretical data, it is recommended that different correlations are used for the mean convective heat transfer dependent on the transition occurring at $\beta_{tr,\omega}$. Its value can be obtained from the empirical correlation (6.22), i.e.

$$\frac{\beta_{tr,\omega}}{[°]} = 60.6 \left(\frac{d}{R}\right)^{1/2} - 21.9 \left(\frac{d}{R}\right) + \exp\left(1.545 \frac{Re_\omega}{Re_u}\right). \tag{8.7}$$

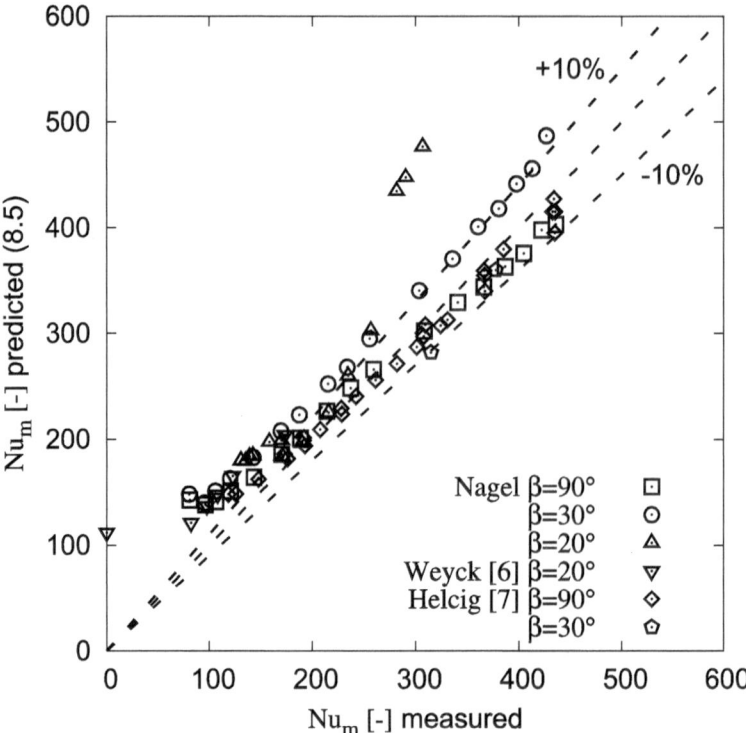

Fig. 8.2 Performance chart for heat transfer correlation (8.5) considering (8.6) for inclined stationary disks $(\beta > \beta_{tr})$

In (8.7), the value for $\beta = \beta_{tr,\omega}$ is expressed in degrees (°). For a large incidence, $\beta > \beta_{tr,\omega}$, the configuration corresponds to an orthogonal rotating disk. Then, the mean heat transfer is given neglecting rotation or turbulent flow by means of the Beg correlation

$$Nu_{m,\omega=0} = 0.63 \ Re_u^{1/2} \quad \text{for} \quad \beta > \beta_{tr,\omega}. \tag{8.8}$$

The range of validity of that laminar correlation can be assumed to be of an order $Re_u < 10^6$ in the case of inflows with moderate or low turbulence intensity. Due to rotation, heat transfer augmentation occurs for the entire range of the rotational Reynolds number $Re_\omega > 0$ but its impact is very small in comparison to the stagnation flow mechanism covered by correlation (8.8). This effect was discussed in more detail in Chap. 4. A practical correlation covering the heat transfer due to rotation can be expressed by means of

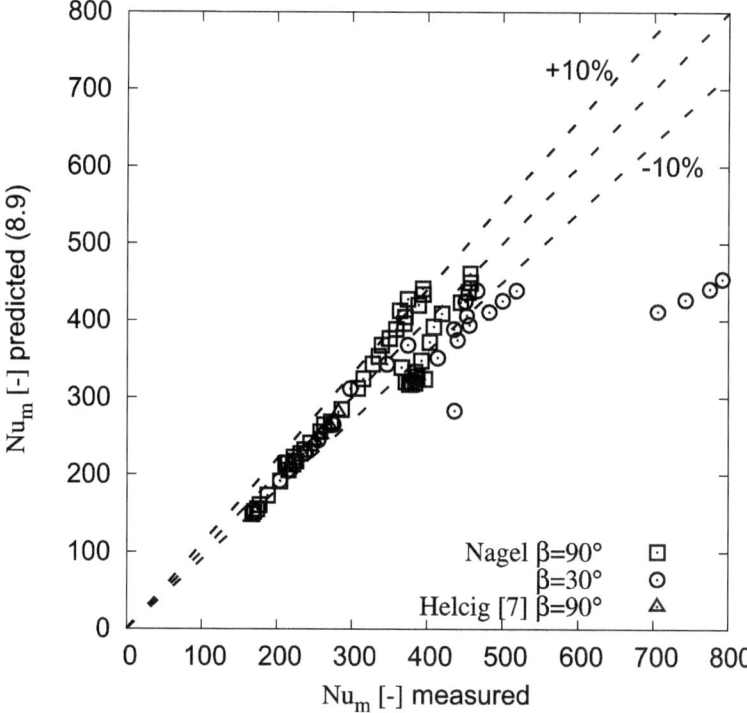

Fig. 8.3 Performance chart for heat transfer correlations (8.7), (8.9), and (8.4) for inclined rotating disks

$$Nu_{\mathrm{m}} = 0.63 \; Re_u^{1/2} \left(1 + 0.21 \left(\frac{Re_\omega}{Re_u} \right)^2 \right)^{1/4} \quad \text{for} \quad \beta > \beta_{\mathrm{tr},\omega}. \qquad (8.9)$$

Correlation (8.9) avoids the rather tricky numerical solution of the correction function f for the self-similar potential flow approach (see Chap. 4 and (4.20)). For a Reynolds number ratio that is not too large, the simple correlation (8.9) yields errors in line with the exact self-similar solution, not exceeding 2 %. For configurations $\beta \leq \beta_{\mathrm{tr},\omega}$, a simple correlation such as (8.9) cannot be given. In fact, the mean heat transfer depends on both Reynolds numbers and incidence. Furthermore, in the vicinity $\beta = \beta_{\mathrm{tr},\omega}$, the onset of transition of the boundary layer flow past the inclined disk from laminar to turbulent is strongly affected by the inflow turbulence level, too. Some data demonstrating the rather complex behavior in this flow and heat transfer regime were provided in Chap. 6. As a practical rule, the mean heat transfer level can be roughly estimated by applying correlation (8.4) derived for a parallel disk. A performance chart of the above simple expressions is given in Fig. 8.3.

8.3   Effect of Prandtl Number

The above experimental and theoretical treatment was restricted to a fixed Prandtl number $Pr = 0.71$ corresponding to dry air at atmospheric conditions. In many applications, other values for the Prandtl number of the involved fluid have to be considered, and it is therefore natural to ask for the effect of Prandtl number on convective mean heat transfer from a rotating disk.

In the case of a free rotating disk without any additional streams of air, Dorfman obtained the most widely known approximate analytical solution for calculating the laminar constant K for gases [8]. His treatment results in a correlation

$$Nu_{\mathrm{m}} = C \cdot Re_\omega^{1/2} \cdot Pr^{1/2} \quad \text{for} \quad Pr \approx 1 \qquad (8.10)$$

for laminar heat transfer from a rotating disk. The value 1/2 for the exponent of the Prandtl number is limited to Prandtl numbers close to 1. It is also in contrast to the asymptotic value obtained for the classic Blasius boundary layer flow past a flat plate, for which an exponent of 1/3 is found [2]. Based on the analogy between heat and mass transfer, results for very high Schmidt numbers suggest instead of correlation (8.11) an asymptotic correlation

$$Nu_{\mathrm{m}} = C \cdot Re_\omega^{1/2} \cdot Pr^{1/3} \quad \text{for} \quad Pr \to \infty. \qquad (8.11)$$

In the literature, some other empirical expressions for the effect of Prandtl number on the mean laminar heat transfer from a free rotating disk are available [3], but no rigorous experimental proof exists so far.

8.4   Outlook to Future Research

For several decades, convective heat transfer from rotating disks has attracted a large number of researchers and engineers. Many important flow and heat transfer features have been successfully investigated, but there are still some major open questions.

One serious lack of knowledge is caused by the absence of experimental heat transfer data for fluids with Prandtl numbers significantly different from $Pr = 0.7$. So far, only air at atmospheric conditions was used in wind tunnel experiments, and it is therefore not clear how the Prandtl number affects the laminar and turbulent heat transfer from a rotating disk. Resulting from this deficit in experimental results, a research program was initiated in 2014 by the *Deutsche Forschungsgemeinschaft DFG*, but, no data had been made available before the publication of this book.

Another important issue is that in experiments only very low Mach numbers have been considered so far. This means that flow and heat transfer effects due to the compressibility of the fluid are not covered by the available experimental (and the majority of theoretical) work. With regard to high-speed turbomachinery or other applications, that lack of knowledge is also relevant, but it will be hard to close due to the intricate experimental effort required in such investigations.

References

1. Lienhard JH, Lienhard JH (2011) A heat transfer textbook. Dover, New York
2. Gesellschaft VDI (2010) VDI heat atlas. Springer, Berlin
3. Shevchuk IV (2009) Convective heat and mass transfer in rotating disk systems. Springer, Berlin
4. aus der Wiesche S (2007) Heat transfer from a rotating disk in a parallel air crossflow. Int J Therm Sci 46:745–754
5. Dennis RW, Newstead C, Ede AJ (1970) The heat transfer from a rotating disc in an air crossflow. In: Proceedings of 4th International Heat Transfer Conference, Versailles, 1970 (paper FC 7.1)
6. Trinkl CM, Bardas U, Weyck A, aus der Wiesche S (2011) Experimental study of the convective heat transfer from a rotating disc subjected to forced air streams. Int J Therm Sci 50:73–80
7. Helcig C, aus der Wiesche S (2013) The effect of the incidence angle on the flow over a rotating disk subjected to forced air streams. In: Proceedings ASME Fluids Engineering Summer Meeting, Incline Village, Nevada (Paper FEDSM2013-16360)
8. Dorfman LA (1963) Hydrodynamic resistance and the heat loss of rotating solids. Oliver & Boyd, Edinburgh

Index

© The Author(s) 2016
S. aus der Wiesche, C. Helcig, *Convective Heat Transfer From Rotating
Disks Subjected To Streams Of Air*, SpringerBriefs in Applied Sciences
and Technology, DOI 10.1007/978-3-319-20167-2